PREFACE

It gives me an immense pleasure to introduce my new book entitled **Text Book of Industrial**

Chemistry The book contains six chapters, each have justification according to the prescribe

syllabus of Addis Ababa Science and Technology University. It has got its own importance in

the field of science, this book will keep the interest of the scientists, academicians and approach

of Industrial Chemistry, main motto of my book is to enhance the knowledge of science for the

next generation of the contemporary world. I have made an effort to bring it in the form of

systematic approach. The book is useful for the both under graduate and post graduate students

as well as for the scientists. The book is very useful to acquire deep knowledge in the field of

science.

I am very grateful to Prof. S. R Niranjana Vice- Chancellor. And also am thankful to Registrar

Prof. Somashekarappa and Prof. D. M Madari Registrar (Evaluation) for their suggestions and

cooperation. I am very thankful to Dr. M.V.N. Ambika Prasad for his support during my PhD

research and thereafter. I am thankful to Dr. Anilkumar R Koppalkar and Dr. Ameena Parveen

for their support, suggestions and cooperation to bring this book to readers, also thankful to

Prof. A Venkataramana, Dean faculty of Sciences and Professor in Department of Chemistry,

Dr. R. L. Raibagkar. Professor. Dept. of PG Studies and Research in Applied Electronics,

Gulbarga University Kalaburagi for their formal or informal support.

I very thankful to my family members, without their cooperation and support I will become a

layman, their nature and attitude supported me to carry out the research in the field of research, so I am very grateful to Parents, brothers and sisters, in laws, and pupils of our family. For their constant help in carrying research activities without their support book in hand would not be possible. I am very thankful to all my students' friends, research scholars of the Dept of History and all my degree college friends all who have formally or informal helped me during course of my research activities.

Dr. Aashish Roy
Assistant Professor
Department of Industrial Chemistry
Adj. Faculty of Center for Nanotechnology
Addis Ababa Science and Technology
University, Addis Ababa.

FOREWORD

It's a matter of immense significance that Dr. Aashish Roy, Assistant Professor, Department of Industrial Chemistry, Addis Ababa Science and Technology University. who has taken this good effort in the field of research activities. He is young, enthusiastic critical thinker and good academician at national and international. In this book, design and manufacturing of industrial compounds are well explained. As a young scholar Dr. Aashish Roy is pursuing his academic and research activities in this part of the region, I bless him to success in his academic, research activities and in his carrier.

DR. Prakash M. Badiger
Historian
Gulbarga University, Kalaburagi-585106.

FOREWORD

It's a matter of immense significance that Dr. Aashish Roy, Assistant Professor, Department of Industrial Chemistry, Addis Ababa Science and Technology University. who has taken this good effort in the field of research activities. He is young, enthusiastic critical thinker and good academician at national and international. In this book, design and manufacturing of industrial compounds are well explained. As a young scholar Dr. Aashish Roy is pursuing his academic and research activities in this part of the region, I bless him to success in his academic, research activities and in his carrier.

Dr. Ameena Parveen
Assistant Professor
Government First Grade College, Gurumatkal
Karnataka, India

CONTENTS

CHAPTER I: COAL AND PETROLEUM PROCESSING

1. Introduction

The first coal age formation began during the carboniferous period, which spanned 360 million up to 290 million years ago. Tectonic movements is movements in the earth's crust which build-up of silt and other sediments together buried swamps and peat bogs, often to great depths. With burial, the plant material was subjected to high temperature and pressures. This leads physical and chemical changes in the vegetation, transforming it in to peat and then into coal.

1.1 Coalification

Coalification is the degree of change undergone by coal as it matures from peat to anthracite. It has an important bearing on coals physical and chemical properties and is referred to as the 'rank' of the coal. The degree of transformation of original plant material to carbon is determined by ranking. The ranks of coals, from those with the least carbon to those with the most carbon are lignite, sub-bituminous, bituminous and anthracite.

Coal deposit quality is determined by:

> ➢ Vegetation types which coal originated
> ➢ Depths of burial
> ➢ Both temperatures and pressures at those depths
> ➢ Duration of time the coal has been forming in deposit

1.1.1 Ranking of coals

As geological processes apply pressure to dead biotic material over time, its metamorphic grade increase successfully continuous,

1. Peat, called a precursor of coal, it serves as fuel in some industrial regions, for example, Ireland and Finland. In dehydrated form, peat is a highly effective 7absorbent for fuel and oil spills on land and water. It used as conditioner for soil to make it more able to retain and slowly release water.

2. Lignite (brown coal): is lowest. It exclusive as for electric power generation compact form of lignite polished and has been used as an ornamental stone since the upper palaeolit.

3. Sub-bituminous coal, properties range from those of lignite to bituminous coal, is used primary fuel for steam-electric power generation and an important source of light aromatic hydrocarbon for the chemical synthesis industry.

4. Bituminous coal is dense cole, usually black, but sometime dark brown, often with

5. Well-defined bands of bright and dull material. Bituminous coal is used as fuel in steam-electric.

6. Power generation, with substantial quantities used for and power application in manufacturing and to make coke.

7. "Steam coal" is a grade between bituminous coal and anthracite. It used as a fuel for steam locomotives. Steam coal sometime known as "sea coal "in the US. Small steam coal (dry small steam nuts or DSSN) used as a fuel for domestic water heating.

8. Anthracite, the highest rank of coal, hard and glossy black coal used primarily for residential and commercial space heating. It divided into metamorphically altered bituminous coal and" petrified oil" as form the deposits in Pennsylvania.

9. Graphite is difficult coals to ignite. It is mostly used in pencils and powdered as lubricant.

1.1.2 Carbonization of coal

The Earth has dense forest in low-lying wetland areas four long time in geologic past. Due to natural processes such as flooding, these forests were buried underneath soil. More soil deposited, they were compressed. As sank deeper and deeper, temperature also rose. Plant matte was protected from biodegradation and oxidation as the process continued usually by mud or acidic water. This trapped the carbon in immense peat bogs that were eventually covered and deeply buried by sediments. Dead vegetation was slowly converted to coal under high pressure and high temperature. Coal contain carbon and the conversion of dead vegetation into coal is called carbonization.

The wide, shallow seas of the Carboniferous Period provided ideal conditions for coal formation, although coal is known from most geological periods. The exception is the coal gap in the Permian–Triassic extinction event, where coal is rare. Coal is known from Precambrian strata, which predate land plants — this coal is presumed to have originated from residues of algae.

1.1.3 Gasification of coal

To produce syngas coal gasification is used by mixture of carbon monoxide (CO) and hydrogen (H_2) gas. Syngas is used to fire gas turbines to produce electricity, but the versatility of syngas also allows it to be converted into transportation fuels, such as gasoline and diesel. Syngas can be converted into methanol, which can be blended into fuel directly process. The hydrogen obtained from gasification that used for powering a hydrogen economy, making ammonia, or upgrading fossil fuels.

Coal is mixed with oxygen and steam while also being heated and pressurized during gasification. During this reaction, during the reaction, oxygen and water molecules oxidize the coal into carbon monoxide (CO), while also releasing hydrogen gas (H_2). Both production of town gas and underground coal mines in gasification process.

$$C_{coal} + O_2 + H_2O \rightarrow H_2 + CO$$

At this stage syngas is collected and routed in to Fischer-Tropsch reaction that wants to produce gasoil from refiner. If hydrogen is the desired end-product, however, the syngas is fed into   the water gas shift reaction, where more hydrogen is liberated

1.1.4. Liquefaction of coal or Hydrogenation of coal or Bergius process

A method of production of liquid hydrocarbons for use as synthetic fuel by hydrogenation of high-volatile bituminous coal at high temperature and pressure is called Bergius process. Thise process was first developed by Friedrich Bergius in 1913.

In 1931 Bergius was awarded the Nobel Prize in Chemistry for his development of high pressure chemistry.

Finally, coal is ground and dried in a stream of hot gas. Dry product is mixed with heavy oil recycled from the process. Typically, catalyst is added to the mixture. Catalyst including tungsten or molybdenum sulphides or nickel oleate and others. On other hand iron sulphides present in the coal may have sufficient catalytic activity for the process which was the original Bergius process. They also considered with respect to global warming, especially if coal hydrogenation(liquefaction) is conducted without carbon carbon capture and storage technologies.

Mixture is pumped in to a reaction and reaction occurs between 400 and 500C* and 20 to 70 Mpa a hydrogen pressure. The reaction produces heavey oils, middle oils, gse oils and gases. The overall reaction can be summarized as follows:

$$_nC + (n- x + n)H_2 \longrightarrow C_nH_{2n-2x+2} \text{ (where x is degree of unsaturation)}$$

The prompt item from the reactor must be balanced out by disregarding it an ordinary hydrotreating impetus. The item stream is high in naphthenes and aromatics, low in paraffins and low in olefins. The distinctive portions can be passed to additionally handling (splitting, changing) to yield manufactured fuel of attractive quality. On the off chance that went through a procedure, for example, Platforming, a large portion of the naphthenes are changed over to aromatics and the recouped hydrogen reused to the procedure. The fluid item from Platforming will contain more than 75% aromatics and has a Research Octane Number (RON) of more than 105.

Generally speaking, around 97% of information carbon nourished straightforwardly to the procedure can be changed over into manufactured fuel. Be that as it may, any carbon utilized in producing hydrogen will be lost as carbon dioxide, so diminishing the general carbon productivity of the procedure.

There is a buildup of lifeless hesitate mixes blended with fiery debris from the coal and impetus. To limit the loss of carbon in the deposit stream, it is important to have a low-slag feed. Commonly the coal ought to be <10% fiery debris by weight. The hydrogen required for the procedure can be additionally delivered from coal or the deposit by steam transforming. A run of the mill hydrogen request is ~80 kg hydrogen per ton of dry, fiery debris free coal. By and large, this procedure is like hydrogenation. The yield is at three levels: substantial oil, center oil, fuel. The center oil is hydrogenated keeping in mind the end goal to get more fuel and the substantial oil is blended with the coal again and the procedure restarts. Along these lines, substantial oil and center oil portions are likewise reused in this procedure.

1.2. Petroleum – Origin, Classification and mining

Petroleum is a normally happening, yellow-to-dark fluid found in topographical developments underneath the Earth's surface, which is generally refined into different kinds of energizes. Segments of petroleum are isolated utilizing a procedure called partial refining.

It comprises of hydrocarbons of different sub-atomic weights and other natural mixes. The name oil covers both normally happening natural unrefined petroleum and oil based commodities that are comprised of refined raw petroleum. A non-renewable energy source, petroleum is formed when vast amounts of dead life forms, typically zooplankton and green growth, are covered underneath sedimentary shake and subjected to both extraordinary warmth and weight.

By alluding to various grounds from two contradicting hypothetical theory, oil starting point arrangement still turn into a captivated subject of researchers' discussions. These hypotheses are abiogenesis and biogenesis. Abiogenesis-inorganic starting point of petroleum, is a most seasoned hypothesis which recommends that petroleum originates from the underneath part of the mantle long time prior before the presence of life on earth (Mendeleev, 1877). The second theory, biotic or natural cause recommends that oil is framed from organic issues, abandoned by extremely old lives. These issues progress toward becoming subjected to high temperature under the nonattendance of oxygen. The last theory, biogenesis is right now acknowledged by numerous individuals because of how it is bolstered by different legitimate grounds while the first is more dubious. Its initial strong precepts lost their fact, particularly when they fall in compression with current science.

1.2.1 Biogenetic origin of petroleum

"Early in 16th century, "a hypothesis of the source of petroleum expressed that it came about because of profound carbon stores that have been around far longer than life on this planet. That hypothesis, of late wound up known as the abiotic petroleum development (AOF) hypothesis, was to a great extent and overlooked until rather as of late when a couple of human some of them researchers restored it and supported it with somewhere in the range of tenets".

As the earth presence is go back to 4.5 billion years, the Abiotic hypothesis is said to happen in that time, before the presence of any type of life. The theory bases on the way that some of reaped hydrocarbons and other related substances have a profound beginning, in reality they are broadly found in the universe. Methane is said to be available in the environment of Jupiter, Saturn, Uranus, on others planets and also moons and shooting stars found in the close planetary system. Russian scientific expert and mineralogist Dimitri Mendeleev and researchers of the age have had an extraordinary impact supporting the theory. "They suggest that abiogenic methane mirrors an inestimable natural legacy that is therefore discharged by the mantle and relocates towards the surface using shortcomings in the hull, for example, plate". As of late in twentieth century, individuals from purported "Russian-Ukrainian School" bolstered the theory by expressing that produced methane polymerizes into higher sub-atomic weight hydrocarbons which results into oil stores, the reality which is likewise convince by finding expanded plenitude of methane gas in the profundity of oil bowl. The fundamentals supporting abiogenic cause of oil are in the accompanying way.

I.   The existence of methane on other planets of solar system, meteors, moons and comets.
II.   The biogenic explanation fails to explain some of hydrocarbon deposit characteristics.

III.   The crude oil distribution of metals fits better with upper serpentinized mantle, primitive mantle and chondrite patterns than the oceanic and the continental crust, and never shows any correlation with sea water.

IV.   The helium and other noble gas association with hydrocarbons.

V.   Deep hydrocarbon seeps.

VI.   Hydrocarbon-rich areas tend to be hydrocarbon-rich at various different levels.

VII.   Some proposed mechanisms of abiogenesis formation of petroleum.

1.2.2 Biogenesis

Biogenetic beginning of oil (Hydrocarbons) recommends that oil originate from quite a while rotting of kicked the bucket life forms, for example, tiny fishes, zooplankton advertisement other type of organic species under a subjection of high temperature. This speculation is right now acknowledged by numerous individuals around the globe and it has numerous suitable supporting grounds which fits well present day sciences. As per that speculation, long time back, the life forms (marine living things, earthbound) kicked the bucket and covered and canvassed by residue in a sedimentary bowl where they experience a moderate and dependable physical and synthetic change which includes procedures, for example, diagenesis and kerogen development.

The more regular perspective of oil development is that it framed when chosen aliquots of biomass from dead living beings were covered in a sedimentary bowl and subjected to diagenesis through delayed introduction to microbial rot taken after by expanding temperatures and weights. Oxygen-poor conditions, created by fatigue of neighborhood oxygen levels by biomass rot and frequently maintained by physical obstructions to oxygen energize, are clear enhancers for fossil natural issue conservation and section into the geosphere. The real natural parts in life are huge, high sub-atomic weight substances and the most safe of these units are safeguarded in dregs, expanded by cross-connecting responses that polymerize and consolidate littler units into the mind boggling system. The high atomic weight sedimentary natural issue is named kerogen from the Greek for "wax previous." It is significant that not the majority of life's natural

7

issue is reflected in kerogen. Indeed, even under moderately ideal conditions less that 1% of the beginning life form, speaking to the most safe compound constituents, might be saved (Demaison and Moore 1980).

The speculation of biotic beginning of oil has numerous conceivable confirmations which can for sure enable researchers to reproduce the generation of oil (unrefined petroleum). Today headways in science, for example, science learning about carbon and its mixes and geography make the theory surely knew and well valuable. The most conceivable confirmation is the attention on the phase of what purported "advancement of hydrocarbons", from peat to anthracite and similarly from green growth to oil.

1.3 Composition

In its strictest sense, oil incorporates just raw petroleum, however in like manner use it incorporates all fluid, vaporous and strong hydrocarbons. Under surface weight and temperature conditions, lighter hydrocarbons methane, ethane, propane and butane happen as gases, while pentane and heavier ones are as fluids or solids. Be that as it may, in an underground oil supply the extents of gas, fluid, and strong rely upon subsurface conditions and on the stage graph of the oil blend.

An oil well delivers prevalently unrefined petroleum, with some gaseous petrol disintegrated in it. Since the weight is bring down at the surface than underground, a portion of the gas will leave arrangement and be recouped (or consumed) as related gas or arrangement gas. A gas well delivers dominatingly petroleum gas. Nonetheless, on the grounds that the underground temperature and weight are higher than at the surface, the gas may contain heavier hydrocarbons, for example, pentane, hexane, and heptane in the vaporous state. At surface conditions these will consolidate out of the gas to frame petroleum gas condensate, regularly abbreviated to condensate. Condensate takes after fuel in appearance and is comparative in creation to some unpredictable light unrefined oils.

The extent of light hydrocarbons in the oil blend changes significantly among various oil fields, extending from as much as 97 percent by weight in the lighter oils to as meager as 50 percent in the heavier oils and bitumens.

The hydrocarbons in raw petroleum are for the most part alkanes, cycloalkanes and different fragrant hydrocarbons while the other natural mixes contain nitrogen, oxygen and sulfur, and follow measures of metals, for example, press, nickel, copper and vanadium. Many oil repositories contain live microscopic organisms. The correct sub-atomic piece fluctuates broadly from development to arrangement yet the extents of compound components change over genuinely slender points of confinement as takes after:

Table 1: Composition by weight of petroleum

Sl.No	Element	Wight percentages
1.	Carbon	83 to 85 %
2.	Hydrogen	10 to 14 %
3.	Nitrogen	0.1 to 2%
4.	Oxygen	0.05 to 1.5 %
5.	Sulfur	0.05 to 6 %
6.	Metal	< 0.1%

1.3.1 Distillation of petroleum

Cruide petroleum is a mind boggling blend of numerous mixes, yet essentially hydrocarbon compound atoms. A blend comprises of at least two components or mixes which are NOT artificially consolidated.

The concoction properties of every substance in the blend is unaltered as the there are no compound bonds between the hydrocarbon atoms. In this way a blend can be isolated effortlessly by physical means case fragmentary refining.

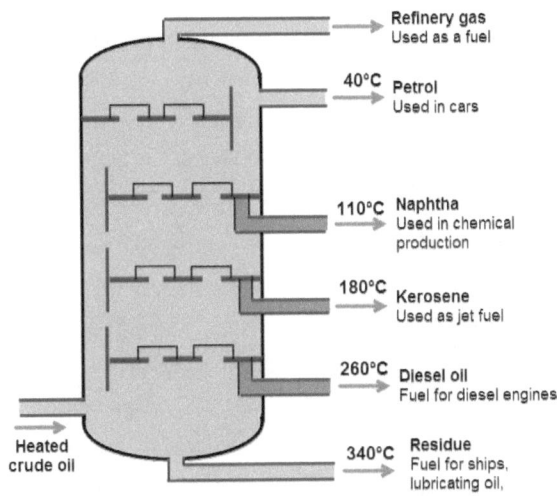

Figure 1.1 show the fractional distillation refinery set up

i. This means crude oil can be separated by physical methods, in this case by fractional distillation, because they have different boiling and condensation points.
   (a) The liquids must also be completely soluble in each other, that is they must all be miscible liquids.
   (b) When the temperature is high enough, the kinetic energy of a particular hydrocarbon molecule will be sufficient for it to escape the intermolecular forces in the liquid and become a gas.
   (c) The intermolecular forces are much weaker than the strong carbon - carbon bonds in the hydrocarbon molecule, so it vaporizes without decomposes.
ii. At the bottom of the fractionating column the crude oil is heated to vaporize it (evaporated or boiled) and the vapor passed into the fractionating column,
   (a) A fractionating column acts in the same way as a fractional distillation apparatus in the school/college laboratory but on an industrial scale!

10

(b) In an oil refinery the fractionating columns are very tall with huge surface area to give the best chance of separating the dozens of hydrocarbons in the crude oil.

iii. This is a continuous process (not a batch process). The fractionating column works continuously with heated–vapourised crude oil piped in at the bottom and the various fractions condensed and constantly tapped off from various levels, each with a different condensation temperature range.

iv. Up the fractioning column the temperature gradually decreases (temperature gradient), so the highest boiling (least volatile) molecules tend to be at the bottom and the lowest boiling (most volatile) hydrocarbons go to the top. The rest of the hydrocarbon molecules then condense out in narrow temperature range i.e. the different fractions condense out in a gradual way from top to bottom depending on their boiling point.

v. In other words the most volatile fraction, i.e. the molecules with the lowest boiling points (shortest hydrocarbon molecules), boil or evaporate off first and go higher up the column and condense out at the higher levels in the fractionating column at the lowest temperature.

vi. The rest of the hydrocarbon molecules separate out according to their boiling/condensation point so that the highest boiling fraction, i.e. the less volatile molecules with higher boiling points (longest hydrocarbon molecules), tend to condense more easily lower down the column, albeit at the higher temperatures.

vii. The process is perhaps more correctly called fractional condensation but it is still referred to as fractional distillation.

viii.   The bigger the molecule, the greater the intermolecular attractive forces between the molecules, so the higher the boiling point or condensation point.

    a) This is an important rule to know since the intermolecular forces (intermolecular bonding) affect the physical properties including melting point and viscosity too, and this has a bearing on how each fraction is used, see above.

    b) Covalent chemical bonds like C–C or C–H are not broken in the process, only the intermolecular force of attraction is weakened to allow the initial evaporation or boiling and this.

ix. The fractions are then further processed to produce fuels and chemical feedstock for the petrochemical industry.

x. These include fuels such as liquified petroleum gas, petrol, diesel oil, kerosine, heavy fuel oil which are all non-renewable fossil fuels, as is methane from natural gas.

1.3.2 Rating of petrol and diesel

Octane rating or octane number is a standard measure of the performance of an engine or aviation fuel. The higher the octane number, the more compression the fuel can withstand before detonating (igniting). In broad terms, fuels with a higher octane rating are used in high performance gasoline engines that require higher compression ratios. In contrast, fuels with lower octane numbers (but higher cetane numbers) are ideal for diesel engines, because diesel engines (also referred to as compression-ignition engines) do not compress the fuel, but rather compress only air and then inject fuel into the air which was heated by compression. Gasoline engines rely on ignition of air and fuel compressed together as a mixture, which is ignited at the end of the compression stroke using spark plugs. Therefore, high compressibility of the fuel matters mainly for gasoline engines. Use of gasoline with lower octane numbers may lead to the problem of engine knocking.

1.3.3 Pre-ignition and knocking

In a normal spark-ignition engine, the air-fuel mixture is heated due to being compressed and is then triggered to burn rapidly by the spark plug. If it is heated (or compressed) too much, it will self-ignite before the ignition system sparks. This causes much higher pressures than engine components are designed for, and can cause a "knocking" or "pinging" sound. Knocking can cause major engine damage if severe.

The octane rating of gasoline is measured in a test engine and is defined by comparison with the mixture of 2,2,4-trimethylpentane (iso-octane) and heptane that would have the same anti-knocking capacity as the fuel under test: the percentage, by volume, of 2,2,4-trimethylpentane in that mixture is the octane number of the fuel. For example, gasoline with the same knocking characteristics as a mixture of 90% iso-octane and 10% heptane would have an octane rating of 90. A rating of 90 does not mean that the gasoline contains just iso-octane and heptane in these proportions but that it has the same detonation resistance properties (generally, gasoline sold for common use never consists solely of iso-octane and heptane; it is a mixture of many hydrocarbons and often other additives). Because some fuels are more knock-resistant than pure iso-octane, the definition has been extended to allow for octane numbers greater than 100.

Octane ratings are not indicators of the energy content of fuels. (See Effects below and Heat of combustion). They are only a measure of the fuel's tendency to burn in a controlled manner, rather than exploding in an uncontrolled manner. Where the octane number is raised by blending in ethanol, energy content per volume is reduced.

1.3.4 Rating Diesel, Understanding Cetane Numbers

Hydrocarbons, a compound comprised of carbon and hydrogen, typically make up the bulk of various petroleums. Cetane is one such hydrocarbon that appears in a colorless fluid form. When it comes under extreme compression, the pressure causes it to ignite quite quickly. Cetane is assigned the base rating of a hundred. Accordingly, it is then used as a measurement to indicate how well other fuels, like diesels or even biodiesels, perform.

Standard gasoline uses a rating system of octane numbers to measure its efficiency. In a similar manner, diesel uses cetane numbers as a measurement of how well it combusts. However, there is a significant difference in what exactly these two ratings measure. For gasoline, the octane number measures how well the substance will resist spontaneously combusting at average temperatures without a helper ignition source. This is known as auto-ignition. Comparatively, the cetane number (or CN for short) tests how long the diesel delays its ignition time after the fuel enters the combustion chamber.

With standard cars, it generally does not benefit the engine if the driver fills it with a gasoline that features an octane rating more than the amount recommended by the manufacturer. Similarly, vehicles that run on diesel do not perform any better by using fuel with a cetane number higher than the recommended amount for that specific engine. Despite this, there is sometimes a misconception that any higher cetane number will result in better engine performance and power.

There are many different factors that are involved in determining the best cetane number for a certain type of engine. This normally includes the physical size of the engine, the way it is designed, how fast it operates, as well as its load variations. A less factor, but one that is counted all the same, is external weather or climate conditions. On the other hand, if an engine is operated with a fuel that features a cetane number lower than the recommended amount, there can be several drawbacks. The vehicle will not operate as smoothly, and the poor operation can result in vibrations as well as extra noises. Additionally, it could create a larger amount of emissions and wear on the engine. In some cases, the driver may even have difficulty starting the engine.

1.3.5 Types of Diesel Fuels and Corresponding Cetane Numbers

Most diesels for standard vehicles and general highway usage normally require a cetane rating that falls between 45 to around 55. The table below outlines the different grades of cetane numbers that correspond with various diesel fuels that are compression ignited.

Table 2 shows the types of diesel.

Types of Diesel	Cetane Numbers
Regular Diesel	48
Premium Diesel	55
Biodiesel (B100)	55
Biodiesel Blend (B20)	50
Synthetic Diesel	55

1.3.6 Cracking, Alkylation, Hydrotreating and Reforming

Cracking is the process whereby complex organic molecules such as kerogens or long chain hydrocarbons are broken down into simpler molecules such as light hydrocarbons, by the breaking of carbon-carbon bonds in the precursors. The rate of cracking and the end products are strongly dependent on the temperature and presence of catalysts. Cracking is the breakdown of a large alkane into smaller, more useful alkanes and alkenes. Simply put, hydrocarbon cracking is the process of breaking a long-chain of hydrocarbons into short ones. This process might require high temperatures and high pressure.

A large number of chemical reactions take place during the cracking process, most of them based on free radicals.

Initiation: In these reactions a single molecule breaks apart into two free radicals. Only a small fraction of the feed molecules actually undergo initiation, but these reactions are necessary to produce the free radicals that drive the rest of the reactions. In steam cracking, initiation usually involves breaking a chemical bond between two carbon atoms, rather than the bond between a carbon and a hydrogen atom.

$$CH_3CH_3 \rightarrow 2\ CH_3\bullet$$

Hydrogen abstraction: In these reactions a free radical removes a hydrogen atom from another molecule, turning the second molecule into a free radical.

$$CH_3\bullet + CH_3CH_3 \rightarrow CH_4 + CH_3CH_2\bullet$$

Radical decomposition: In these reactions a free radical breaks apart into two molecules, one an alkene, the other a free radical. This is the process that results in alkene products.

$$CH_3CH_2\bullet \rightarrow CH_2{=}CH_2 + H$$

Radical addition: In these reactions, the reverse of radical decomposition reactions, a radical reacts with an alkene to form a single, larger free radical. These processes are involved in forming the aromatic products that result when heavier feedstocks are used.

$$CH_3CH_2\bullet + CH_2\!=\!CH_2 \rightarrow CH_3CH_2CH_2CH_2\bullet$$

Termination: In these reactions two free radicals react with each other to produce products that are not free radicals. Two common forms of termination are recombination, where the two radicals combine to form one larger molecule, and disproportionation, where one radical transfers a hydrogen atom to the other, giving an alkene and an alkane.

$$CH_3\bullet + CH_3CH_2\bullet \rightarrow CH_3CH_2CH_3$$
$$CH_3CH_2\bullet + CH_3CH_2\bullet \rightarrow CH_2\!=\!CH_2 + CH_3CH_3$$

Example: Cracking butane

There are three places where a butane molecule (CH_3-CH_2-CH_2-CH_3) might be split. Each has a distinct likelihood:

➢ 48%: break at the CH_3-CH_2 bond.

CH_3* / $*CH_2$-CH_2-CH_3

Ultimately this produces an alkane and an **alkene**: CH_4 + CH_2=CH-CH_3

➢ 38%: break at a CH_2-CH_2 bond.

CH_3-CH_2* / $*CH_2$-CH_3

Ultimately this produces an alkane and an alkene of different types: CH_3-CH_3 + CH_2=CH_2

➢ 14%: break at a terminal C-H bond

H/CH_2-CH_2-CH_2-CH_3

Ultimately this produces an alkene and hydrogen gas: CH_2=CH-CH_2-CH_3 + H_2

1.3.7 Types of cracking

(a) Thermal cracking

Breaking down large molecules by heating at high temperature and pressure is termed as thermal cracking. Thermal cracking is further classified into the following classes.

Liquid phase thermal cracking process: The higher boiling fractions e.g., fuel oil, lubricating oil are converted into low boiling fractions by heating the liquids at a temperature of 750 K, under a pressure of about 10 atmosphere.

Vapor phase thermal cracking process: Low boiling fraction e.g., kerosene is cracked in the vapor phase at a temperature of about 875 K and under a pressure of 3 atmosphere.

(b)   Catalytic cracking

Higher hydrocarbons can also be cracked at lower temperature (600 - 650 K) and lower pressure (2 atm) in the presence of a suitable catalyst. Catalytic cracking produces gasoline of higher octane number and therefore this method is used for obtaining better quality gasoline. A typical catalyst used for this purpose is a mixture of silica (SiO_2), 4 parts; alumina (Al_2O_3), 1 part, and manganese-dioxide (MnO_2), 1 part.

(c) Steam cracking

Here, higher hydrocarbons are mixed with steam in their vapor phase and heated for a short duration to about 900°C, and cooled rapidly. This process is suitable for obtaining lower unsaturated hydrocarbons.

1.4 Applications of cracking

The most important products obtained in straight-run refining are petrol, diesel and kerosene. The demand for these products outstrips that obtained during refining of the petroleum. Conversely, high-boiling fractions find lesser use. So,

In petroleum industry, the cracking of less useful high boiling fractions is done to increase the yield of low boiling (lower molecular mass) fractions, such as gasoline.

Cracking always yields low boiling alkenes as the by-products. These unsaturated hydrocarbons are called petrochemicals, that form a variety of useful compounds such as, polyethylene etc.

1.4.1 An Alkylation is one of the conversion processes used in petroleum refineries. It is used to convert isobutane and low-molecular-weight alkenes (primarily a mixture of propene and butene) into alkylate, a high octane gasoline component. The process occurs in the presence of a strong acting acid such as sulfuric acid or hydrofluoric acid (HF) as catalyst. Depending on the acid used, the unit takes the name of SAAU (Sulphuric Acid Alkylation Unit) or HFAU (Hydrofluoric Acid Alkylation Unit).

For example, when isobutene undergoes protonation in presence of propene to form Alkylate (2,4 – Dimethyl Pentane).

An example alkylation reaction

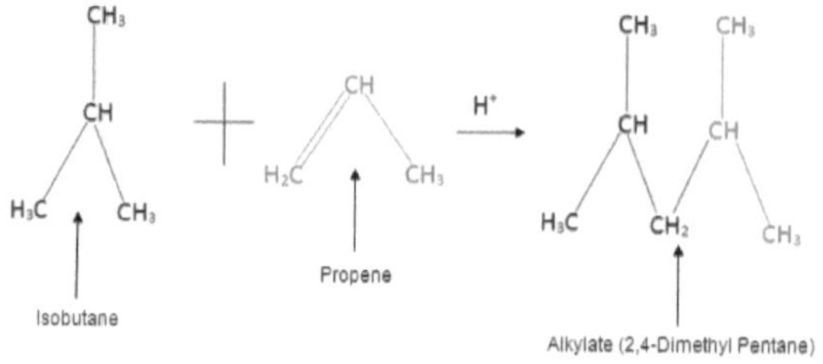

CH₃

CH

H₃C   CH₃

Isobutane

+

CH

H₂C   CH₃

Propene

H⁺ →

CH₃   CH₃

CH   CH

H₃C   CH₂   CH₃

Alkylate (2,4-Dimethyl Pentane)

1.42. Hydrotreating

The type and amount of impurities to be removed by catalytic hydrotreating in a petroleum distillate can vary substantially depending on the type and source of the feed. In general, light feeds (e.g., naphtha) contain very little and few types of impurities, while heavy feeds (e.g., residua) possess most of the heavy compounds present in a crude oil. Apart from having a high concentration of heavy compounds, the impurities in heavy feeds are more complex and refractory (i.e., difficult to react) than those present in light feeds. That is why hydrotreating of light distillates is conducted at lower reaction severity, whereas heavy oils require higher reaction pressures and temperatures.

The reactions occurring during catalytic hydrotreating can be classified in two types: hydrogenolysis and hydrogenation. In hydrogenolysis a carbon-heteroatom single bond undergoes "lysis" by hydrogen. The heteroatom is any atom other than hydrogen or carbon present in petroleum, such as sulfur, nitrogen, oxygen, and metals. In hydrogenation, hydrogen is added to the molecule without cleaving bonds. The principal hydrogenolysis and hydroge-nation reactions in catalytic hydrotreating are described below.

1.4.3 Hydrogenolysis reactions

(a) **Hydrodesulfurization (HDS)** is the removal of organic sulfur compounds from a petroleum fraction and conversion to hydrogen sulfide (H2S). Sulfur removal difficulty increases in the following order: paraffins < naphthenes < aromatics. The type of sulfur compounds can be classified as mercaptans, sulfides, disulfides, thiophenes, benzothiophenes, dibenzothiophenes, and substituted dibenzothiophenes. The ease of removal of these sulfur compounds is in the same order, the mercaptans being the easiest to remove and dibenzothiophenes the most difficult.

(b) **Hydrodenitrogenation (HDN)** is the removal of organic nitrogen compounds and conversion to ammonia (NH3). Removal of nitrogen requires more severe reaction conditions than does HDS. The molecular complexity (five- and six-membered aromatic ring structures), the quantity, and the difficulty of nitrogen-containing molecules to be removed increase with increasing boiling range of the distillate. Nitrogen compounds can be basic or nonbasic. Pyridines and saturated heterocyclic ring compounds (indoline, hexahydrocarabazole) are generally basic, whereas pyrroles are nonbasic.

(c) **Hydrodeoxygenation (HDO)** is the removal of organic oxygen compounds and conversion to water. Similar to HDS and HDN, lower- molecular-weight oxygen compounds are easily converted, while higher-molecular-weight oxygen can be difficult to remove. Phenol is one of the most difficult oxygen compounds to convert.

(d) **Hydrodemetallization (HDM)** is the removal of organometals and conversion to the respective metal sulfides. Nickel and vanadium being the most common metals present in petroleum, hydrodemetallization is frequently subdivided into hydrodeniquelization (HDNi) and hydrodevanadization (HDV). Once metal sulfides are formed, they are deposited on the catalyst and contribute to irreversible deactivation.

20

Asphaltenes can undergo both types of reactions (hydrocracking and hydrogenation) depending on reaction conditions. At relatively low or moderate temperatures, the reaction is more hydrogenation dominated during hydrocracking of heavy residue; however, at high temperatures, hydrocracking is more prominent. The overall conversion of asphaltenes is called hydrodeas-phaltenization (HDAsp). Examples of some typical reactions occurring during catalytic hydrotreating are presented in below scheme,

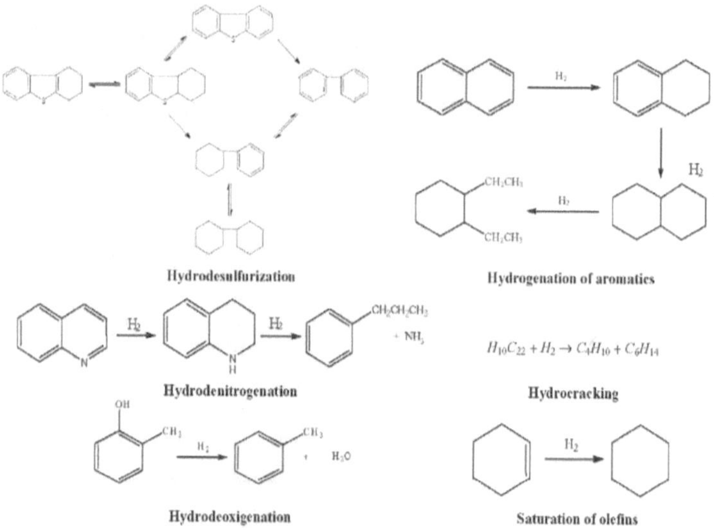

Hydrodesulfurization

Hydrogenation of aromatics

$$H_{10}C_{22} + H_2 \rightarrow C_4H_{10} + C_6H_{14}$$

Hydrodenitrogenation

Hydrocracking

Hydrodeoxigenation

Saturation of olefins

1.5 Reforming

The change of straight chain hydrocarbon into stretched chain hydrocarbon is called improving of oil. By the way toward transforming, the octane number of a fuel is made strides. The way toward changing is done within the sight of impetus

The n-alkanes consume in inner ignition motor with blast and create thumping however expanded chain hydrocarbons consume easily. Improving is a procedure like breaking, which changes over n-alkanes into spread alkanes.

The transforming procedure includes three separate synergist reactors, every one occurring under precisely controlled temperature and weight levels. Naphtha is blended with hydrogen and encouraged through every reactor chamber in arrangement. Extra hydrogen framed by the synergist reactors is recuperated and put to use in resulting transforming and in different procedures all through the refinery. Alternate results of improving are light gases and a high-octane gas mixing part called reformate.

The octane rating of reformate is imperative since it influences the octane rating of the gas you purchase at the pump. By controlling the temperature and stream rate of the reformer, refinery administrators can build the octane rating of reformate, yet that additionally has the impact of creating less reformate. The invert is likewise valid: If interest for high-octane gas is lower, the reformer can be acclimated to create more reformate with a lower octane rating.

CHAPTER II: PETROCHEMICAL PRODUCTS

2. Introduction

At the point when individuals tune in to the term petrochemical as often as possible accept of plastic or different composite item got from oil. Either that or they accept of oil based solvents like those utilized in paints and coatings. In the strictest logical sense, petrochemicals are an arrangement of unmistakable substance compound romds, which can be produced using oil, gaseous petrol, coal or different sources. The expansive quantities of petrochemicals, be that as it may, are gotten from oil or petroleum gas. Oil and flammable gas are utilized as feedstock's (i.e., the primary crude material utilized in the fabricate of an item) to make roughly 99% of petrochemicals.

Different petrochemicals are produced utilizing extraordinary temperature (more than 1500 ^{0}F) and weights (more than 1000 psi) This procedure requires gigantic measures of vitality and complex building. As a result of the extraordinary working circumstance, vitality utilization represents huge segment of the aggregate cost of generation. As vitality costs rise (expands), the cost of working together additionally incresis chance to shabby and tried and true vitality sources, (for example, flammable gas) is basic for guaranteeing the petrochemical business stays focused in an undeniably overall commercial center.

Petrochemical are the essential building square of natural science. The fundamental case of petrochemicals are: Ethylene, Propylene, Butadiene, Benzene, Toluene, Xylene such sorts of petrochemicaals originates from other an extensive number of synthetics, which are classified, "petrochemical subordinates" or basically, "subsidiaries." The subordinates are assembled in view of number of step requires to change over the essential compound into the new subsidiary. For instance, it makes one move to change over ethylene to acetaldehyde; along these lines acetaldehyde can be viewed as a first-subsidiary of ethylene.

2.1 Chemical conversion for manufacturing of petrochemicals

The adjacent diagram schematically depicts the major hydrocarbon sources used in producing petrochemicals are:

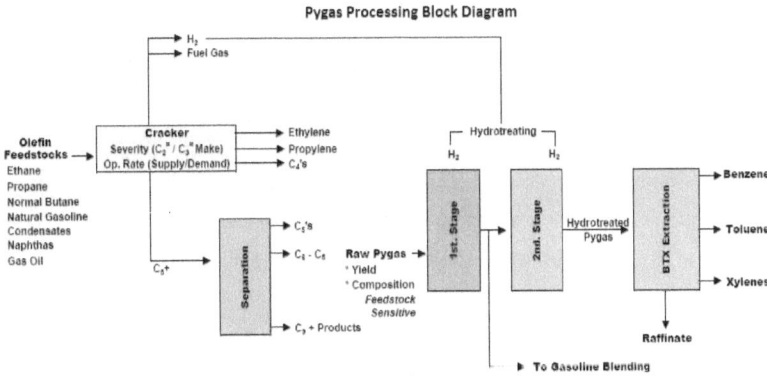

Figure 2.1 shows the cracking of olefin into petroleum products

Methane, ethane, propane and butanes: Obtained primarily from natural gas processing plants.

➢ Naphthal obtained from petroleum refineries.

- Benzene, toluene and xylenes, as a whole referred to as BTX and primarily obtained from petroleum refineries by extraction from reformate produced in catalytic reformers.
- Gas obtained from petroleum refineries.

Methane and BTX are used directly as feed stocks for producing petrochemicals. However, the ethane, propane, butanes, naphtha and gas oil serve as optional feed stocks for steam-assisted thermal cracking plants referred to as steam crackers that produce these intermediate petrochemical feed stocks:

- Ethylene
- Propylene
- Butenes and butadiene
- Benzene

Steam crackers are not to be confused with steam reforming plants used to produce hydrogen and ammonia.

2.1.1 Methane (CH_4)

Methane is dismal, unscented combustible gas which is the fundamental constituent of petroleum gas and the least complex individual from the alkane arrangement. It is likewise a side-effect in all gas streams from handling rough oils. It is lighter than air.

Table 1 shows selected physical properties of C1–C4 paraffinic hydrocarbon gases.

Name	Formula	Specific gravity	Boiling point °C	Calorific Value Btu/ft^3
Methane	CH_4	0.554	-161.5	1.009
Ethane	CH_3CH_3	1.049	-88.6	1.800
Propane	$CH_3CH_2CH_3$	1.562	-42.1	2.300
n-Butane	$CH_3(CH_2)_2CH_3$	-.579	-0.5	3.262

| Isobutene | $(CH_3)_2CHCH_3$ | 0.557 | -11.1 | 3.253 |

As a chemical compound, methane isn't exceptionally receptive. It doesn't respond with acids or bases under ordinary conditions. It responds, be that as it may, with a predetermined number of reagents, for example, oxygen and chlorine under particular conditions. For instance, it is in part oxidized with a restricted measure of oxygen to a carbon monoxide-hydrogen blend at high temperatures in nearness of an impetus. The blend (combination gas) methane isn't extremely responsive as a substance compound under typical conditions. It doesn't respond with acids or bases anyway It responds, with a predetermined number of reagents, for example, oxygen and chlorine under particular conditions. For instance, it is somewhat oxidized with a constrained measure of oxygen to a carbon monoxide-hydrogen blend at high temperatures in nearness of an impetus. The blend (amalgamation gas) is an essential building obstruct for some synthetic substances. Methane is basically utilized as a perfect fuel gas. Around one million BTU are acquired by consuming 1,000 ft3 of dry petroleum gas (methane). It is likewise a critical hotspot for carbon dark. Methane might be melted under high weights and low temperatures. Liquefaction of gaseous petrol (methane), enables its transportation to long separations through cryogenic tankers.

2.1.2 Ethylene ($CH_2=CH_2$)

Ethylene (ethene), the main individual from the alkenes, is a drab gas with a sweet scent. It is combustible hydrocarbon gas of the alkene arrangement, happening in petroleum gas and coal gas ethane. It is marginally solvent in water and liquor.

Expansion of chlorine to ethylene produces ethylene dichloride (1,2-dichloroethane), which is broken to vinyl chloride. Vinyl chloride is a critical plastic forerunner. Ethylene is additionally a functioning alkylating specialist. Alkylation of benzene with ethylene produces ethyl benzene, which is dehydrogenated to styrene. Styrene is a monomer utilized in the produce of numerous business polymers and copolymers.

Styrene is unsaturated fluid hydrocarbon acquired as an oil side-effect and used to make plastic and tars. It is a monomer utilized in the produce of numerous business polymers and copolymers. Ethylene can be polymerized to Catalytic oxidation of ethylene produces ethylene oxide, which is hydrolyzed to ethylene glycol. Ethylene glycol is a monomer for the generation of engineered strands.

2.2 Steam cracking of ethylene

The fundamental course to produce light olefins, particularly ethylene, is the steam splitting of hydrocarbons. The feedstock's for steam breaking units go from light paraffinic hydrocarbon gases to different oil divisions and deposits. The most essential route for creating light olefins, particularly ethylene, is the steam breaking of hydrocarbons. The splitting responses are chiefly bond breaking, and a generous measure of vitality is expected to drive the response toward olefin creation. The easiest paraffin (alkane) and the most generally utilized feedstock for delivering ethylene is ethane. As said before, ethane is gotten from petroleum gas fluids. Ethane is combustible hydrocarbon gas of the alkane's arrangement, display in oil and flammable gas. Ethane is the easiest paraffin (alkanes) and the most generally utilized feedstock for delivering ethylene and it is acquired from gaseous petrol fluids. Splitting ethane can be envisioned as a free extreme dehydrogenation response, where hydrogen is a co-product:

$$CH_3-CH_3 \longrightarrow CH_2=CH_2 + H2 \; \Delta H_{590°C} = +143 \; KJ$$

The response is very endothermic, so it is favored at higher temperatures and lower weights. Super warmed steam is utilized to lessen the halfway weight of the responding hydrocarbons' (in this response, ethane). Superheated steam additionally lessens carbon stores that are framed by the pyrolysis of hydrocarbons at high temperatures. For instance, pyrolysis of ethane produces carbon and hydrogen:

$$CH_3-CH_3 \longrightarrow 2C + 3H_2$$

Ethylenes can also pyrolyse in the same way. Additionally, the presence of steam as a diluent reduces the hydrocarbons' chances of being in contact with the reactor tube-wall. Deposits reduce heat transfer through the reactor tubes, but steam reduces this effect by reacting with the carbon deposits (steam reforming reaction).

$$C + H_2O \longrightarrow CO + H_2$$

Many side reactions occur when ethane is cracked. A probable sequence of reactions between ethylene and a formed methyl or an ethyl free radical could be represented:

$$CH_2=CH_2 + CH_3 \longrightarrow CH_3-CH=CH_2 + H$$

$$CH_2=CH_2 + CH_3CH_2 \longrightarrow CH_3CH_2CH_2CH_2 \longrightarrow CH_3CH_2CH=CH_2 + H$$

Propene and 1-butene, respectively, are produced in this free radical reaction. Higher hydrocarbons found in steam cracking products are probably formed through similar reactions.

2.3 Preparation of Acetylene: The Caving Helmet

Description: The calcium carbide lamp of a Caving Helmet is shown and demonstrated.

Principle: Acetylene gas produced by the reaction of Water with calcium carbide which can be used to light a lamp on a caving helmet.

Materials:

- Caving Helmet with calcium carbide lamp

- Calcium carbide
- Pin to clear lamp outlet for the gas
- Water
- Matches
- Forceps

Safety:

Make sure fire extinguisher is handy. Wear gloves and safety goggles.

Procedure:

If you have not used a calcium carbide lamp before. Obtain assistance from Matthew Nance or Tom Hacker.

1. Put enough carbide chips to cover the bottom of the container. Close the container tightly.
2. At the off position, add water about 10 minutes before you want to light the lamp.
3. Turn to on position and wait. It takes about 10 minutes for the water to start dripping into the lower compartment where the calcium carbide is.
4. After about 10 minutes light the lamp.

Tips:

1. use pin to clear outlet

2. use match to light the lamp if striker does not work

Clean-up:

Let lamp sit in fume hood until reaction has completed (no acetylene odor). Empty contents onto aluminum foil and let stand until no acetylene odor. Wash contents of lamp down the drain. Insert pin in outlet to keep it clear.

Background:

Reaction:

$$CaC_2 + 2 H_2O \rightarrow H\text{-}C \equiv C\text{-}H + Ca(OH)_2$$

calcium carbide   acetylene   hydrated lime

2.4 Manufacture of Ethylene oxide (Epoxyethane)

Ethene is blended with air or oxygen and disregarded an impetus (finely separated silver on a latent help, for example, alumina) at 520-550 K and under 15-20 airs weight. Two responses, halfway and finish oxidation, happen at the same time at the impetus surface. A little measure of 1,2-dichloroethane is added to the response blend which diminishes the undesirable side response to carbon dioxide and steam. Living arrangement time in the reactor is 1-4 seconds.

Silver is extraordinary as an impetus for this response however the component isn't clear. The selectivity presently being accomplished is more than 90%. As the impetus ages, selectivity diminishes. The lifetime of an impetus is in the scope of 2-5 years.

Chemical reaction,

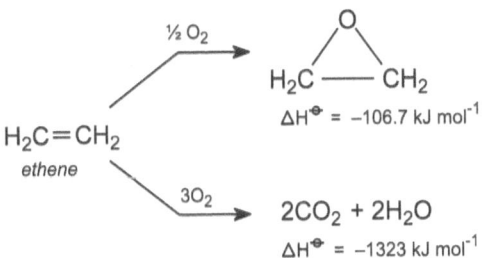

Ethylene oxide, properly called oxirane by IUPAC, is the organic compound with the formula C_2H_4O. It is a cyclic ether. (A cyclic ether consists of an alkane with an oxygen atom bonded to two carbon atoms of the alkane, forming a ring.) Ethylene oxide is a colorless flammable gas at room temperature, with a faintly sweet odor; it is the simplest epoxide: a three-membered ring consisting of one oxygen atom and two carbon atoms. Because of its special molecular structure, ethylene oxide easily participates in addition reactions; e.g., opening its ring and thus easily polymerizing. Ethylene oxide is isomeric with acetaldehyde and with vinyl alcohol.

Despite the fact that it is an imperative crude material with differing applications, including the make of items like polysorbate 20 andpolyethylene glycol (PEG) that are regularly more successful and less lethal than elective materials, ethylene oxide itself is an extremely unsafe substance. At room temperature it is a combustible, cancer-causing, mutagenic, bothering, and sedative gas, with a misleadingly wonderful smell.

The concoction reactivity that is in charge of huge numbers of ethylene oxide's perils has likewise made it a key modern synthetic. Albeit excessively hazardous for coordinate family unit utilize and for the most part new to shoppers, ethylene oxide is utilized mechanically to make numerous customer items and in addition non-buyer synthetics and intermediates. Ethylene oxide is vital or basic to the creation of cleansers, thickeners, solvents, plastics, and different natural synthetic substances, for example, ethylene glycol, ethanolamines, straightforward and complex glycols, polyglycol ethers and different mixes. As a toxic substance gas that leaves no deposit on things it contacts, unadulterated ethylene oxide is a disinfectant that is generally utilized in healing centers and the medicinal hardware industry to supplant steam in the cleansing of warmth delicate instruments and gear, for example, dispensable plastic syringes.

2.5 Manufacture of Acrylonitrile

Acrylonitrile is a natural compound with the equation CH2CHCN. It is a boring unstable fluid, albeit business tests can be yellow because of debasements. As far as its atomic structure, it comprises of a vinyl gather connected to a nitrile. It is a vital monomer for the produce of helpful plastics, for example, polyacrylonitrile. It is responsive and dangerous at low measurements.

Acrylonitrile is delivered by synergist ammoxidation of propylene, otherwise called the SOHIO procedure. In 2002, world creation limit was assessed at 5 million tons for every year. Acetonitrile and hydrogen cyanide are critical results that are recouped available to be purchased. Indeed, the 2008– 2009 acetonitrile deficiency was caused by a decline popular for acrylonitrile.

$$2CH_3\text{-}CH=CH_2 + 2NH_3 + 3O_2 \rightarrow 2CH_2=CH\text{-}C\equiv N + 6H_2O$$

In the SOHIO procedure, propylene, smelling salts, and air (oxidizer) are gone through a fluidized bed reactor containing the impetus at 400– 510 °C and 50– 200 kPag. The reactants go through the reactor just once, before being extinguished in watery sulfuric corrosive. Abundance propylene, carbon monoxide, carbon dioxide, and dinitrogen that don't disintegrate are vented specifically to the climate, or are burned. The fluid arrangement comprises of acrylonitrile, acetonitrile, hydrocyanic corrosive, and ammonium sulfate (from abundance smelling salts). A recuperation section evacuates mass water, and acrylonitrile and acetonitrile are isolated by refining. Verifiably, one of the principal effective impetuses was bismuth phosphomolybdate upheld on silica as a heterogeneous impetus. Promote upgrades have since been made.

2.6 Manufacture of Dimethyl terephthalate

Dimethyl terephthalate (DMT) is an organic compound with the formula $C_6H_4(CO_2CH_3)_2$. It is the diester formed from terephthalic acid and methanol. It is a white solid that melts to give a distillable colourless liquid. Dimethyl terephthalate (DMT) has been produced in a number of ways. Conventionally and still of commercial value is the direct esterification of terephthalic acid. Alternatively, it can be prepared by alternating oxidation and methyl-esterification steps from p-xylene via methylptoluate

32

The common method for the production of DMT from para-xylene (PX) and methanol consists of four major steps: oxidation, esterification, distillation, and crystallization. A mixture of p-xylene (PX) and p-Toluic ester (PT) is oxidized with air in the presence of a heavy metal catalyst (Co/Mn). The acid mixture resulting from the oxidation is esterified with methanol (MeOH, CH_3OH) to produce a mixture of esters. The crude ester mixture is distilled to remove all the heavy boilers and residue produced; the lighter esters are recycled to the oxidation section. The raw DMT is then sent to the crystallization section to remove DMT isomers, residual acids and aromatic aldehydes.

Esterification of the resulting acid with methanol results in Methyl p-toluate at 250^0C and 2500 kPa and subsequent oxidation and esterification of Methyl p-toluate yields Dimethyl terephthalate (DMT) as shown in the below reaction:

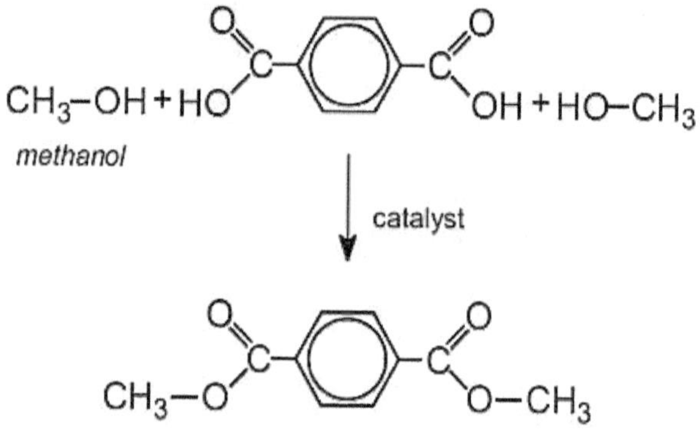

2.7 Dimethyl Terephthalate (DMT) production through Direct Esterification

If highly impure Terephthalic acid is available, DMT can be made in a separate process by esterification with methanol to dimethyl terephthalate which is then purified by distillation:

$$C_8H_6O_4 \text{ (TPA)} + 2CH_3OH \text{ (Methanol)} \rightarrow C_{10}H_{10}O_4 \text{ (DMT)} + 2\,H_2O$$

at presence of o-xylene at 250–300 ^0C.

CHAPTER III: BASIC ORGANIC PRODUCTS

3. Introduction

Environmentally Identified Products (EIPs) refers to food products that are described as organic or sustainable, or were grown using Integrated Pest Management (IPM), or in other respects are viewed as having a relatively less negative impact on the environment than directly competing products. Also, organic synthesis industry belongs to the most important branches of the modern chemical industry.

Manufacturing of chemical products necessary for the production of polymers, chemical fibers, resins, dyes, pesticides for agriculture and other compounds without which the progress of many industrial branched is impossible.

In the middle of the last century practically all organic compounds were obtained from raw plant and animal materials. Today carbon containing products (natural and associated gases, oil-refinery gases, coke gases, crude oil, etc.) are the main source for the production of various organic compounds. The organic compounds obtained via synthetic methods may fully substituent the natural materials. Moreover, they have the valuable properties which are absent in natural compounds. Also, the synthetic methods do not exclude the possibility of using products of plant and animal origin as raw materials at the same time..

The main raw material at the beginning organic synthesis development was BLACK COAL. From coal tar pitch obtained coking, drugs, dyes, explosives, etc. were produced. The peculiarity of the industrial processes in the end of XIX century was the use of liquid phases, low temperatures and pressures as well as non-catalytic systems.

Black coal as the main raw material for the organic synthesis industry dominated in Western European countries in 20-50th of the past century. Among main directions of its application were:

- From coal tar pitch, production of aromatic hydrocarbons

- From syngas, production of methanol

- From syngas via Fisher-Tropsh method,And production of motor fuels via coal hydrogenation.

- From calcium carbide and obtaining of acetic aldehyde and acid, vinylchloride, vinylacetate, acrylonitrile, acrylic acid and other products based on acetylene, production of acetylene.

The rapid growth of organic synthesis industry was also provided by numerous scientific-technical achievements at the beginning of XIX century. The most important were the development of thermal cracking and oil pyrolysis, natural gas processing, production of calcium carbide. These processes supplied the organic synthesis industry by olefins and acetylene, i.e. by main primary products. Using catalytic systems accelerated and simplified the most of multistage methods. The development of distillation and sorption theory gives the possibility for the development of clear rectification methods for polycomponent mixtures.

3.1 Manufacture of Methanol

Many millions of tons of Methanol are produced by the Chemical Industry every year. Methanol is the lowest member of a group of organic chemicals that belong to the Alcohol 'family'. Methanol is extremely poisonous to humans, even when ingested in very small quantities. Methanol is a basic 'building block' for the production of other chemical products such as Plastics, Paints and Man-Made Fibers. It has also found a large use in the production of Gasoline Fuel additives. The production of methanol usually consists of three basic steps independent of feedstock material; synthesis gas preparation, methanol synthesis and methanol purification.

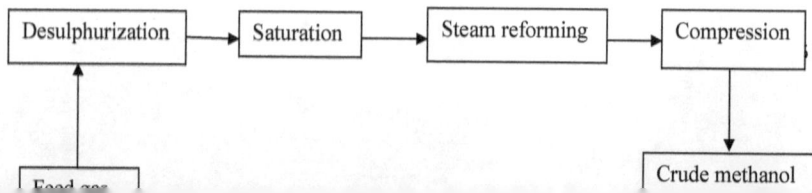

Figure 3.1 Schematic diagram of catalytic process to prepare methanol

Natural gases such as methane used to prepare the methanol by catalytic process. When feed gas erupted from the earth crust contains different particles such as sulphur, calcium and small amount of iron particles. The feed gases under goes desulphurization in presence of zinc oxide as catalyst to obtain sulphur free feed gases then passes into saturation tower where the water vapour sprinkled from the top. The calcium and iron particles absorb water molecules become heavy and settled down at the bottom. Saturated methane gases subjected to steam reforming in presence of nickel catalyst at 850 °C to form Syngas. It is the mixture of H_2, CO and CO_2 formed by steam reforming process compressed at the 100 bar for injecting in to the methanol synthesis loop as shown in scheme 3.1.

$$CO + 2H_2 \rightleftharpoons CH_3OH \qquad \Delta H298K= -21.7 \, kcal/mol \qquad (1)$$

$$CO_2 + 3H_2 \rightleftharpoons CH_3OH + H_2O \quad \Delta H298K=-11.9 \, kcal/mol \qquad (2)$$

$$Reverse \, WGS \; CO_2 + H_2 \rightleftharpoons CO+H_2O \quad \Delta H298K= 9.8 \, kcal/mol \qquad (3)$$

All three equations are reversible and thus the process conditions regarding temperature, pressure and synthesis gas mixture are important to control. It can also be noted that equation (1) and (2) are exothermic, i.e. the processes produce heat and require cooling. Some heat is normally recovered and used for other parts of the synthesis.

While it was initially trusted that the fundamental procedure that create methanol was the response between carbon dioxide and hydrogen (condition 1) it is currently comprehended that carbon dioxide is similarly as essential in the union procedure. CO_2 even used to be cleaned from the reactant blend however a scrubber disappointment at Imperial Chemical Industries (ICI) with a subsequent increment of methanol generation demonstrated that CO_2 was dynamic and critical in the response. Ensuing investigations have demonstrated that it is fundamentally the CO_2 that is changed over to methanol while CO go about as a lessening specialist for oxygen at the surface of the impetus. Condition (3) portrays the turnaround water gas move response that produces carbon monoxide from carbon dioxide and hydrogen. The carbon monoxide at that point responds with hydrogen to create methanol (condition 2). Condition 2 is quite the total of (1) and (3). To combine methanol, not exclusively is a particular H_2/CO proportion of 2 in the orchestrated gas required yet additionally a $(H_2-CO_2)/(CO+CO_2)$ proportion, called stoichiometric number, equivalent to or somewhat over 2.

3.2 Manufacture of Isopropanol

Significant business techniques for the modern generation of isopropanol include synthetic blend from propylene and water. Likewise, the hydrogenation of side-effect $CH_3)_2CO$ is polished financially for low volume isopropanol generation. Other engineered strategies have been examination in the research facility yet not completely created to business scale. These incorporate aging of specific sugars, oxidation of propane, and hydrolysis of isopropyl acetic acid derivation. For the motivations behind this report, center is given to business creation strategies right now by and by, with joining of pertinent bits of knowledge and advancements from the autonomous writing. Specialized data is aggregated underneath for the three monetarily applicable engineered forms, and in addition advancements in the free writing for the fermentative generation of isopropanol. Industrial production of isopropanol involves chemical synthesis from propylene and water. In addition, the hydrogenation of by-product acetone is practiced commercially for low volume isopropanol production. Technical information is compiled below for the three commercially relevant synthetic processes, as well as developments in the independent literature for the fermentative production of isopropanol.

3.2.1 Indirect Hydration

The indirect hydration, also known as the sulfuric acid process, was the only process used worldwide from 1920 until ICI developed an industrial direct hydration process in 1951. Propylene ($CH_3CH=CH_2$) and water are the chemical feedstocks for isopropanol formation in the indirect process. Indirect hydration can tolerate lower purity streams of propylene from refineries. In the indirect hydration process, C3-feedstock streams from crude oil refinery off-gases containing 40–60 percent propylene ($CH_3CH=CH_2$) are subjected to sulfuric acid (H_2SO_4) to generate both isopropyl hydrogen sulfate [$(CH_3)_2CHOSO_3H$] and diisopropyl sulfate $[((CH_3)_2CHO)_2SO_2]$. These sulfate intermediates are then hydrolyzed with water to generate the desired product, isopropanol, and release sulfuric acid for further reaction cycles. The reaction mixture is neutralized using sodium hydroxide (NaOH) and distilled to afford pure isopropanol. Diisopropyl ether $[((CH_3)_2CH)_2O]$ is the principal by-product formed via reaction of the intermediate sulfate esters with isopropanol, and is generally recycled back to the reactor for hydrolysis to isopropanol. Minor by-products (≤ 2 percent) include acetone, carbonaceous material, and polymers of propylene. See chemical equations below for step one (esterification) and step two (hydrolysis) in the indirect hydration process for isopropanol production in below step 1 & 2.

Step 1 Esterification

$$CH_3CH=CH_2 + H_2SO_4 \leftrightarrow (CH_3)_2CHOSO_3H$$

$$(CH_3)_2CHOSO_3H + CH_3CH = CH_2 \leftrightarrow ((CH_3)_2CHO)_2SO_2$$

Step 2 Hydrolysis

$$(CH_3)_2CHOSO_3H + H_2O \leftrightarrow (CH_3)_2CHOH + H_2SO_4$$

$$((CH_3)_2CHO)_2SO_2 + 2H_2O \leftrightarrow 2 (CH_3)_2CHOH + H_2SO_4$$

3.2.2 Direct Hydration

The direct hydration process addressed many of the early problems associated with the indirect hydration method, including equipment corrosion from concentrated sulfuric acid, high energy costs, and air pollution. However, high purity propylene feedstock is required for this process. Direct hydration is predominantly employed in Europe for industrial isopropanol production.

The acid-catalyzed direct hydration of propylene ($CH_3CH=CH_2$) to form isopropanol $[(CH_3)_2CHOH]$ generally resembles the preparation of ethanol (CH_3CH_2OH) from ethylene ($H_2C=CH_2$). Direct hydrations are conducted using high pressures and low temperatures over an acidic fixed-bed catalyst, which pushes the exothermic (heat releasing) equilibrium reaction toward the formation of isopropanol. Three versions of the direct hydration process are practiced commercially today for 306 isopropanol formation. One method feeds a mixture of propylene gas (92 percent purity) and liquid water to the top of a fixed bed reactor containing a sulfonated polystyrene ion-exchange resin catalyst and allows it to trickle downward. Another direct method reacts propylene (95 percent purity) and water (both gas and liquid phase) over a reduced tungsten oxide catalyst. The final method uses medium to high pressures 310 of high purity propylene (~99 percent) with a tungsten oxide – silicon dioxide (WO_3 – SiO_2) catalyst or a phosphoric acid catalyst supported on SiO_2. The phosphoric acid/SiO_2 process is commercially developed.

$$CH_3CH=CH_2 + H_2O \ catalyst \rightarrow (CH_3)_2CHOH$$

3.3 Manufacture of Formaldehyde

The procedure starts by blending of vaporized methanol and air preceding entering the reactors. Inside the warmth exchanger reactor, the feed is gone through the metal oxide impetus filled tubes where warm is expelled from the exothermic response to the outside of the tubes. Short tubes (1 – 1.5 m) and a shell width 2.5 m is the normal plan of average reactors. The base item leaving the reactors is cooled and gone to the safeguard. The piece of formaldehyde in the safeguard outlet is controlled by the measure of water expansion. A nearly sans methanol item can be accomplished on this procedure outline. The upside of this procedure over the silver based impetus is the nonappearance of the refining section to isolate unreacted methanol and formaldehyde item. It likewise has a life expectancy of 12 to year and a half, bigger than the bit impetus. Nonetheless, the hindrance of this procedure configuration is the requirement for essentially expansive gear to oblige the expanded stream of gases (3 times bigger) contrasted with the first silver impetus process plan as sshown in figure 3.2. This expansion in hardware measuring conflicts with monetary prospect behind the outline costs.

Kinetic information for the methanol oxidation reaction: The methanol compressed along with the air or oxygen in reactor in presence off silver catalyst to form formaldehyde.

$$CH_3OH + \frac{1}{2}O_2 \rightarrow HCHO + H_2O$$

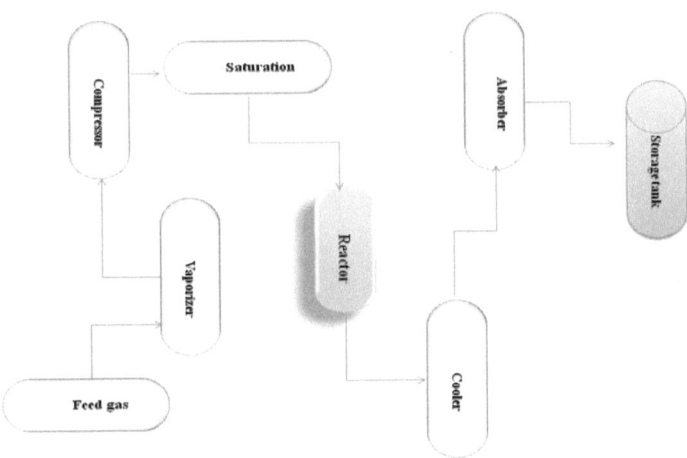

Figure 3.2 Schematic diagram of catalytic process to prepare formaldehyde

3.4 Manufacture of Acetaldehyde

The main difference between the silver process for acetaldehyde production and the one for formaldehyde production is that two distillation columns are needed to produce acetaldehyde. The following section will describe a silver process for acetaldehyde production from ethanol. Air and preheated ethanol goes into a saturator. The air leaving is saturated with ethanol and overheated before entering the reactor.

The ethanol mixed with acidic water in the ratio of 95: 1 introduced in ethanol column compressed 100 bar and passed in saturation tower. The vaporized ethanol sprinkled with water to remove the heavy inorganic and dust particles later introduce in reactor in presence of silver catalyst it covert in to acetaldehyde and produce water as by produce which can be removed in absorption column to obtain pure acetaldehyde as figure 3.3.

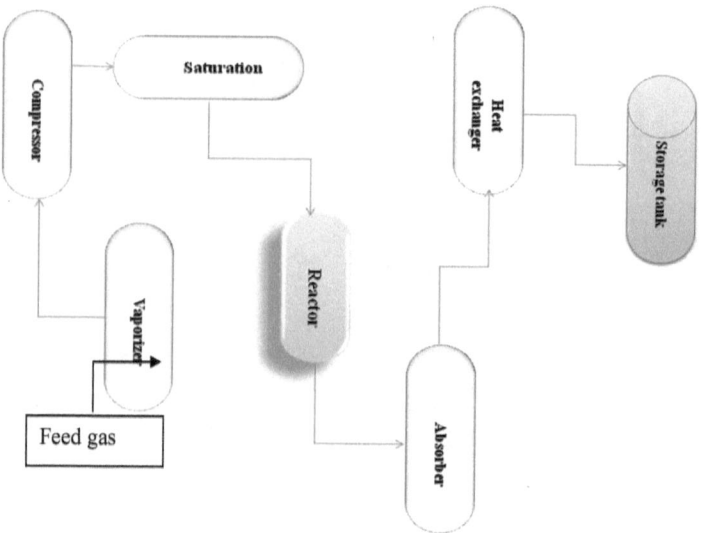

Figure 3.3 Schematic diagram of catalytic process to prepare formaldehyde

The main reactions that take place are:

$C_2H_5OH + \frac{1}{2} O_2 \rightarrow CH_3CHO + H_2O$

$C_2H_5OH \rightarrow CH_3CHO + H_2$

The by-products are formed according to following reactions:

$C_2H_5OH + O_2 \rightarrow CH_3COOH + H_2O$

$C_2H_5OH + \frac{1}{2} O_2 \rightarrow CH_4 + CO + H_2O$

$$C_2H_5OH + 2O_2 \rightarrow 2CO_2 + 3H_2O$$

The overall reaction is exothermic and the temperature in the reactor is 550 °C. The products are cooled instantly and the heat of reaction is used to produce steam. Further cooling is done before the products enter the absorber. The bottom stream is sent to the distillation column where acetaldehyde is removed in the top. The bottom from the distillation column is sent to an ethanol column to recover unconverted ethanol.

3.5 Manufacture of Acetic Acid

The process is carried out in a continuous stirred tank reactor. In this process methanol and CO are reacted in presence of Rhodium complex catalyst to give acetic acid. The liquid feed to the reactor consists of mixture of methanol feed stream used to scrub the effluent gas from the reactor. A recycle stream containing mostly methyl acetate, methyl iodide and recycled catalyst are introduced at the top and CO is spared from the bottom of the reactor. A small amount of methyl iodide and rhodium catalyst will be added to make-up any loss.

The reaction is exothermic and its heat of reaction is 138 KJ/mol. Pressure in the reactor is 3MPa and temperature is around 190^0 C are maintained. All the gaseous effluent from the reactor is cooled to about 10^0C in a water cooled exchanger and condensed in a condenser. The liquid stream from the condenser goes to a high pressure separator to separate the un-reacted CO, CO_2 and H_2. Bottom liquid stream from the high pressure separator goes to low pressure separator for separating the methyl iodide, methyl acetate and mixed with the gas effluent from the high pressure separator. These gases are scrubbed in a scrubber with methanol to recover the methyl iodide and methyl acetate before leaves the system. The liquid from the scrubber is sent to reactor surge tank from there it goes to reactor as shown in figure 3.4.

The liquid from the reactor is removed and the pressure is let down and introduced into the light ends distillation column. Here the low boiling compounds are separated from the acetic acid and other less volatile components such as catalyst. These low boiling components contain methyl acetate and methyl iodide and they are recycled to the reactor.

Acetic acid, other high boiling compounds and rhodium component mixture is sent to catalyst recovery column where catalyst is separated as bottom product and recycled to reactor. The top product contains acetic acid and water is sent to dehydration tower where water is separated from the acetic acid. The acetic acid from the dehydration unit consists small amount of propionic acid to remove it and improve the concentration it is sent to heavier ends distillation column all the propionic acid is removed as bottom product. The top product contains acetic acid with 99% purity.

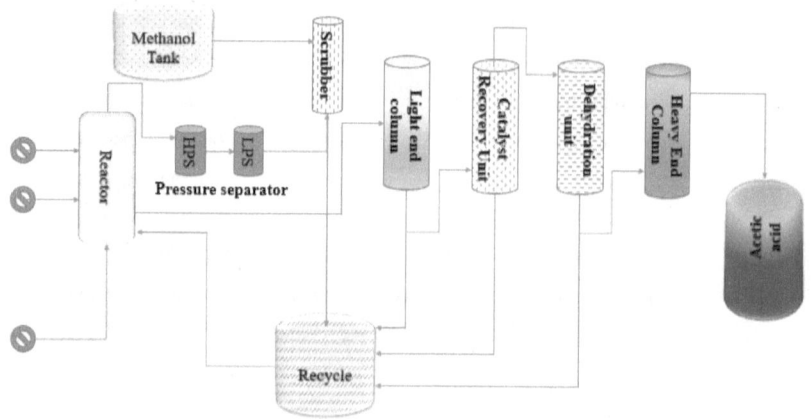

Figure 3. 4 Schematic diagram of catalytic process to prepare acetic acid

1. Oxidation of acetaldehyde:

- Acetaldehyde can be oxidized by the oxygen in the air to produce acetic acid.

  $2CH_3CHO + O_2 \rightarrow 2CH_3COOH$

2. Liquid phase oxidation of butane:

- Liquefied butane is combusted by air in use of various inorganic ions such as those to manganese, cobalt and chromium, peroxides type and then break down to generate acetic acid.

$$2C_4H_{10} + 5O_2 \rightarrow 4\ CH_3COOH + 2\ H_2O$$

3. Methanol Carbonylation

- In this process methanol CH_3OH and carbon-monoxide CO, chemically react for the production of acetic acid according as per the equation.

$$CH_3OH + CO \rightarrow CH_3COOH$$

3.6 Manufacture of Phenol

The vast majority of phenol is made by the cumene process. The process has three stages:

a) Production of cumene

b) Conversion of cumene to cumene hydroperoxide

c) Decomposition of cumene hydroperoxide

a) Production of cumene

Cumene is the name often given to (1-methylethyl)benzene (isopropylbenzene). It is produced by the reaction of benzene and propene, using an acid catalyst; this is an example of a Friedel-Crafts reaction:

$$\text{C}_6\text{H}_6 + CH_3CH{=}CH_2 \xrightarrow[H^+]{\text{catalyst}} \text{C}_6\text{H}_5{-}\overset{\overset{\displaystyle H}{|}}{\underset{\underset{\displaystyle CH_3}{}}{C}}{-}CH_3$$

b) Conversion of cumene to cumene hydroperoxide

Cumene is then oxidized with air to give the hydroperoxide (Figure 3). The reaction is autocatalyzed by cumene hydroperoxide. The overall reaction can be represented as:

$$H_3C{-}\overset{H}{\underset{C_6H_5}{C}}{-}CH_3 \xrightarrow{O_2} H_3C{-}\overset{O_2H}{\underset{C_6H_5}{C}}{-}CH_3 \qquad \Delta H^\oplus = -117 \text{ kJ mol}^{-1}$$

The reaction takes place at temperatures between 350-390 K and 1-7 atm pressure, the latter to retain the system in the liquid phase.

c) Decomposition of cumene hydroperoxide

Finally, the hydroperoxide is mixed with sulfuric acid at 313-373 K to give, after neutralisation, phenol and propanone. This reaction when carried out with small amounts of sulfuric acid (500 ppm by mass) is termed homogeneous cleavage:

$$H_3C{-}\overset{O_2H}{\underset{C_6H_5}{C}}{-}CH_3 \xrightarrow{H^+} C_6H_5OH + H_3C{-}\overset{O}{\overset{\|}{C}}{-}CH_3 \qquad \Delta H^\oplus = -252 \text{ kJ mol}^{-1}$$

The products are separated by distillation, in up to six columns. Product yield is 85-87%, based on benzene.

3.7 Manufacture of Styrene

The development of commercial processes for the manufacture of styrene based on the dehydrogenation of ethylbenzene was achieved in the 1930s. Styrene is a colorless liquid with a distinctive, sweetish odor. Some physical properties of styrene are summarized on the right. Vapor pressure is a key property in the de- sign of styrene distillation equipment.

Ethylbenzene is almost exclusively (>99 %) used as an intermediate for the manufacture of styrene monomer. 50 % of the world's benzene production is consumed for this purpose. Less than 1 % of the ethylbenzene produced is used as a paint solvent or as an intermediate for other chemicals. Currently, almost all ethylbenzene is produced commercially by alkylating benzene with ethylene, primarily via two routes: in the liquid phase with aluminum chloride catalyst (Friedel-Crafts reaction), or in the vapor phase with a fixed bed of either a Lewis acid or a synthetic zeolite catalyst developed by Mobil Corporation.

3.7.1 Oxidation Process

One of the most notable oxidation processes is Halcon international's process to produce styrene from propylene oxide. In this technique, ethyl benzene is oxidized to hydro peroxide as follows:

$$C_6H_5CH_2CH_3 + O_2 \rightarrow C_6H_5CH(OOH)CH_3$$

The reaction takes place in the liquid phase with air bubbling through the liquid, and no catalyst is required. However, since hydro peroxides are unstable compounds, exposure to high temperature must be minimized to reduce the rate of decomposition. Fewer by-products are formed from the decomposition if the reaction temperature is gradually reduced during the course of reaction, i.e., from 135-160 ^{0}C during the first half of reaction to 125-155 ^{0}C during the second half. The reaction is more selective to the production of by-product acids, when it is carried out at constant temperature than when the temperature is gradually reduced. In practice, the temperature is reduced by means of a series of reactors, each of which is maintained at a progressively lower temperature. The pressure required for the reaction is not critical; 800-1500 kPa is sufficient to maintain the reactants at liquid phase.

CHAPTER IV: SUCROSE INDUSTRY

4. Introduction

Sugarcane is comprehensively ordered into three assortments early, general and unapproved. Stick is sowed amid February and October consistently. The main seed development is known as the plant and consequent development subsequent to collecting from the stem is known as Ratoon. The early assortment has more sugar content than the general assortment. Each agriculturist inside the order zone of the Mill is given a schedule, which reveals to him when he can expect a Mill Supply Ticket (Purchy), against which he will convey the sugarcane.

At that point gathers the stick and transports it either in a bullock truck or tractor trolley to the plant. Stick is likewise purchased at the plant's own focuses inside the summon region. This stick is then transported in trucks or through rail to the plant. Stick is measured utilizing an electronic measure connect and emptied into stick bearers. It is then arranged for processing by blades and shredders. Sugarcane juice is then extricated by squeezing the readied stick through factories.

4.1 Manufacture of cane sugar

There are a number of steps in producing raw sugar from cane:

The industrial process for raw sugar manufacturing from cane involves the application of several processes to transform sugarcane juice into crystals and clean them naturally from impurities that can harm the organism. The manufacturing process consists of the following sub-processes:

a) Harvesting

b) Cane preparation (cutting and shredding cane to prepare it for juice extraction)

c) Juice extraction (two technologies are in common use; milling or diffusion)

d) Juice clarification (remove suspended solids from the juice, typically mud, waxes, fibres)

e) Juice evaporation (to concentrate the juice to a thick syrup of about 65°brix)

f) Crystallisation

g) Centrifugation (separation of the sugar crystals from the mother liquor, most done by centrifugal machines).

h) Sugar drying

i) Packaging and delivery

By applying the above manufacturing process either brown or raw sugar will be produced. White sugar can be produced by performing sulphitation (treating the juice with sulfur dioxide) to remove different coloring impurities during juice clarification process. Sugar refining process carried out to produce the white sugar crystal from raw sugar.

4.1.1 Cane harvesting

Sugarcane takes about seven months to mature in tropical area and about 12-22 months in subtropical area. At this time fields of sugarcane are tasted for sucrose, and the most mature fields are harvested first. From this maturity period, cane possesses its maximum sucrose content within 14-18 months. If it is harvested before this month its sucrose content becomes minimum and harvesting after 18 months results in degradation of sucrose contents. Harvesting of sugar is done primarily by machine, although in some state it is also done by hand.

4.1.2 Cane preparation

The process of reducing the size of sugarcane into small pieces which is suitable for subsequent process of extraction is known as cane preparation. Preparation of the cane has three main objectives;

- ✓ To increase capacity by increasing the bulk density of the feed.
- ✓ To assist extraction at the mills by breaking down the structure of the cane.
- ✓ To render the juice more readly available for the action of imbibition by breaking and opening cane cells.

In cane preparation the knifing and shredding of cane takes place to rapture the juice storage celles prior to juice extraction.

4.1.2 Juice extraction

Juice extraction is the process of squazing maximium amount of juice from cane. This cabe done by performing two major techniques; *Diffusion and milling.*

4.1.2 (a) Juice extraction by milling

i) Milling is the procedure by which weight is connected to arranged stick to remove however much squeeze as could reasonably be expected from it. Amid the processing procedure, the stick billets are destroyed and smashed through a few processing units to break the cells and concentrate juice from the sinewy material; Imbibition water is additionally added at the plants to remove the most extreme juice from the stick. Processing trains normally have four, five or six plants in tandems.

ii) Imbibition is the procedure in which the high temp water or squeeze is put on the bagasse to blend with and weaken the juice present in the bagasse. The water so utilized is imbibition water. The separated juice is sent to the procedure, while the sinewy material left in the wake of processing which is known as conclusive bagasse is sent to the evaporator to be utilized as fuel for the age of steam. The significant goals of processing are;

iii) To extract the greatest possible amount of sucrose from sugarcane.

iv) To make the final bagasse as dry as possible so that it will burn readily in the boiler.

4.1.2 (b). Juice extraction by diffusion

Sugarcane diffusion is the recovery of sugars from cane tissue using imbibition water by liquid extraction, as opposed to mechanical expression of the juices in conventional milling practice. In diffusion process, the shredded cane is transferred into the diffuser and mixed with hot water. The hot water percolates through the bed of the cane and removes the sucrose from the cane. This diluted juice is then collected in a compartment under the bed of cane and is pumped to a point a little closer to the feed end of the diffuser and this dilute juice is allowed to percolate through the bed of cane. The extraction of sucrose takes place due to the difference in concentration of sucrose between the cane and the diluted juice. Since the concentration of the sucrose in the cane is higher than that in the juice, sucrose diffuses from the cane to the juice, and this process is repeated for 12-15 times.

4.1.3 Juice clarification

Juice heating: the raw juice collected from miller or diffuser is impure and turbid. Therefore, three heating process (1st, 2nd, and 3rd heating) are carried out for various objectives.
First heating: The juice is heated from 25 to 75. The aims of 1st heating are:

i) To destroy microorganism and enzymes preventing loss of sucrose by microbiological

    activity,

ii) To accelerate rate of reaction and for reaction optimization,

iii) To coagulate organic constituents like proteins.

Sugar stick juice has a pH of around 4.0 to 4.5 which is very acidic. Calcium hydroxide, otherwise called Milk of lime or limewater, is added to the stick juice to change its pH to 7. The lime keeps sucrose's rot into glucose and fructose. The limed juice is then warmed to a temperature over its breaking point. The superheated limed juice is then permitted to glimmer to its immersion temperature: this procedure accelerates contaminations which get held up in the calcium carbonate precious stones. The flashed juice is then exchanged to an illumination tank which enables the suspended solids to settle. The supernatant, known as tidy juice is drawn up of the clarifier and sent to the evaporators.

The main objectives of juice clarification are:

i)   To separate soluble and insoluble matter that can precipitate
ii) To reduce color and turbidity of the juice
iii) To produce clear juice of correct PH of 7.0
iv) To kill or inactivate microorganisms in juice by heat treatment

4.1.4 Sulphitation is an elucidation procedure in which illumination is finished by warmth and sulfur dioxide gas. The blended juice warmed to 75 C°, more lime slurry is added to the crude juice and the abundance of lime is killed by sulphur dioxide (SO_2) gas.

The treated juice is warmed to 98 and partition of flocculated non-sugar happens in nonstop pioneer. From the pilgrim the unmistakable juice proceeds all the while and the undercurrent mud is sifted in nonstop drum channel and the filtrate reused to the deliberate crude juice.

Generally, sulphitation of juice has the following advantages.

i)   Rapid setting of juice
ii) Fast boiling massecuite
iii) Fast crystallization of sucrose
iv) Marked improvement of sugar color

4.1.5 Second heating

The sulphited mixed juice is heated to about 98 – 115 °C. Therefore, the pressure in the juice will be above atmospheric pressure and it is left to flash out its air bubbles that would retard setting of impurities in the clarifier. When the juice is exposed to atmospheric pressure in a flash tank, it loses its excess heat and drops to compatible temperature to atmospheric pressure.

Generally, the purpose of second heating is to increase reaction, to remove unwanted gases and to prevent inversion.

4.1.6 Flash tank

The flashing process will get rid of air or gas bubbles contained in the juice. The treated juice fro the second heating, which is under the pressure provided by sulphitated juice is pumped tangentially into the flash tank. The aim of flashing is to remove bubble gas and also to increase the rate of settling; otherwise they would prevent settling.

4.1.7 Flocculants

It is material of high molecular weight such as poly electrolytes. They are synthetic water soluble, polyacrylamide powders and chemical which are used to increase rate of settling.

The main objectives of flocculants in juice clarification are;

➢ Increase settling rate
➢ Reduce mud volume
➢ Decrease pol in cake
➢ Increase the clarity of juice

Clear juice is introduced to clear juice tank and muddy juice is sent to continuous rotatory vacuum filter for further processing.

4.1.8 Sedimentation (Dorr clarifier):

After the juice is treated with lime and Sulphur between PH= 7-7.2, it is boiled at the temperature of 98, then made to pass through the flash tank and enter in the dorr clarifier where separation by precipitation is done.

Dorr clarifier is an extensive vessel which comprises of four superimposed compartments with feed chamber at the best and each shaping a complete clarifier autonomous of the other. Dorr clarifier is utilized for detachment of sloppy juice from clear squeeze. The mud settles at the base of the clarifier and the reasonable juice stay at the best. The illuminated clear.juice goes into evaporator and the mud send to channel cake.

4.1.9 Rotatory vacuum filter (OC filter): It is a filter composed of hallow drum rotating about a horizontal axis, which is partly submerged in the muddy juice that to be filtered. Generally rotatory filter (OC filter) is used to separate filtrate from the filter cake.

4.2 Juice evaporating: The clarified juice is concentrated in a multiple-effect evaporator to make syrup of about 60 percent sucrose by weight. The reasonable juice that is put away in clear squeeze thank is pumped to the vanishing first body through clear squeeze radiator (third warming). The cleared up juice contains around 85% water, 66% of which is dissipated in a progression of vacuum bubbling cells of various impact evaporators' hence keeping away from demolition of sugar caused by bubbling of juice over direct fire or by steam taking all things together. This additionally results in better efficiency.

These evaporators are orchestrated in an arrangement and the juice is bubbled so that each succeeding evaporator has a higher vacuum and accordingly, bubbles at a lower temperature. The juice along these lines turns out to be increasingly moved and in the last evaporator transforms sugar into syrup containing around 60 - 65%solids and the rest water.

The evaporator station in natural sweetener make regularly delivers syrup with around 65% solids and 35%water. Following dissipation, the syrup is elucidated by including lime, phosphoric corrosive, and a polymer hairy, circulated air through, and sifted in the clarifier. From the clarifier, the syrup goes to the vacuum prospects.

4.3 Crystallisation and centrifuging: This syrup is additionally amassed under vacuum in a vacuum bubbling dish until the point that it winds up supersaturated, finely ground sugar precious stones suspended in liquor are brought into the vacuum container as seed gems around which sucrose is stored and these gems at that point develop in measure until the point when they are prepared to be released. Various bubbling plans are conceivable, the most generally utilized bubbling plan is the three-bubbling plan. This technique heats up the sugar mixers in three phases, called A-, B-and C-.

A clump compose sugar axis isolates the sugar precious stones from the mother alcohol. These rotators have a limit of up to 2,200 kilograms (4,900 lb) per cycle. The sugar from the axes is dried and cooled and after that put away in a storehouse or specifically gathered into packs for shipment.

The mother liquor from the first crystallization step (A-product) is again crystallized in vacuum pans and then passed through continuous sugar centrifuges. The mother-liquor is again crystallized in vacuum pans. Due to the low purity the evapo-crystallization alone is not sufficient to exhaust molasses, and so the so-called massecuite (French for "boiled mass") is passed through cooling crystallizers until a temperature of approx. 45 °C (113 °F) is reached. Then the massecuite is re-heated in order to reduce its viscosity and then purged in the C-produced centrifugals. The run-off from the C-centrifugals is called molasses. The spun-off sugar from the B-product and C-product centrifuges is re-melted, filtered and added to the syrup coming from the evaporator station.

4.4 Drying and packing

In the axis procedure, condensate water is utilized to wash sugar, which results in a stickiness somewhere in the range of 0.3% and 0.6%; along these lines, it is important to go it through the drying procedure to achieve levels between 0.2% for crude sugar and 0.03% for white sugar. Fare crude sugar goes specifically from the dryer to the capacity distribution centers. In the distribution centers, it is stacked in trucks that vehicle the same to the transportation port.

4.5 Back-end refineries: Some cane sugar mills have so-called back-end refineries. In this case, a portion of the raw sugar produced in the mill is directly converted to refined sugar with a higher purity for local consumption, exportation, or bottling companies. Wastage is used for heat generation in the sugar mills.

4.6 Manufacture of sucrose from Beet Root

4.6.1 Reception and cleaning:

The harvested beet root is first transported into the factory for sucrose production. Each load is weighed and sampled before it gets tipped onto the reception area, typically a "flat pad" of concrete, where it is moved into large heaps. After the beets are received, they are washed and cleaned to remove different extraneous materials impurities.

4.6.2 Diffusion: After reception at the processing plant, the beet roots are washed, mechanically sliced into thin strips called cossettes, and passed to a machine called a diffuser to extract the sugar content into a water solution.

Diffusers are long vessels of numerous meters in which the beet cuts go one way while boiling water goes the other way. The development may either be caused by a pivoting screw or the entire turning unit, and the water and cossettes travel through inward chambers. The three basic outlines of diffuser are the level pivoting slanted screw or vertical screw "Tower". Current pinnacle extraction plants have a handling limit of up to 17,000 metric tons (16,700 long tons; 18,700 short tons) every day. A less-normal outline utilizes a moving belt of cossettes, with water pumped onto the highest point of the belt and poured through. In all cases, the stream rates of cossettes and water are in the proportion one to two. Normally, cossettes take around a hour and a half to go through the diffuser, the water just 45 minutes. These countercurrent trade strategies separate more sugar from the cossettes utilizing less water than if they simply sat in a boiling water tank. The fluid leaving the diffuser is called crude juice. The shade of crude juice fluctuates from dark to a dull red relying upon the measure of oxidation, which is itself subject to diffuser outline.

The utilized cossettes, or mash, leave the diffuser at around 95% dampness, yet low sucrose content. Utilizing screw presses, the wet mash is then pushed down to 75% dampness. This recoups extra sucrose in the fluid squeezed out of the mash, and lessens the vitality expected to dry the mash. The squeezed mash is dried and sold as creature feed, while the fluid squeezed out of the mash is joined with the crude juice, or all the more regularly brought into the diffuser at the suitable point in the countercurrent procedure. The last result, vinasse, is utilized as manure or development substrate for yeast societies.

Amid dispersion, a part of the sucrose separates into rearrange sugars. These can experience promote breakdown into acids. These breakdown items are misfortunes of sucrose, as well as have thump on impacts diminishing the last yield of handled sugar from the manufacturing plant. To restrain (thermophilic) bacterial activity, the feed water might be dosed with formaldehyde and control of the feed water pH is likewise polished. Endeavors at working dissemination under basic conditions have been made, however the procedure has demonstrated tricky. The enhanced sucrose extraction in the diffuser is counterbalanced by handling issues in the following stages.

4.6.3 Carbonatation: Carbonatation is a procedure which removes impurities from raw juice before it undergoes crystallization. First, the juice is mixed with hot milk of lime (a suspension of calcium hydroxide in water). This treatment precipitates a number of impurities, including multivalent anions such as sulfate, phosphate, citrate and oxalate, which precipitate as their calcium salts and large organic molecules such as proteins, saponins and pectins, which aggregate in the presence of multivalent cations. In addition, the alkaline conditions convert the simple sugars, glucose and fructose, along with the amino acid glutamine, to chemically stable carboxylic acids. Left untreated, these sugars and amines would eventually frustrate crystallization of the sucrose.

Next, carbon dioxide is bubbled through the alkaline sugar solution, precipitating the lime as calcium carbonate (chalk). The chalk particles entrap some impurities and absorb others. A recycling process builds up the size of chalk particles and a natural flocculation occurs where the heavy particles settle out in tanks (clarifiers). A final addition of more carbon dioxide precipitates more calcium from solution; this is filtered off, leaving a cleaner, golden light-brown sugar solution called thin juice.

In the juice purging stage, non-sucrose pollutions in the crude juice are evacuated with the goal that the unadulterated sucrose can be solidified. Initially, the juice goes through screens to expel any little cossette particles. At that point the juice is warmed to 80 °C to 85 °C (176 °C to 185 °C) and continues to the principal carbonation tank. In a few procedures, the juice from the screen goes through a pre-limer, warmer, and principle limer before the main carbonation tank. In the main carbonation tank, drain of lime [$Ca(OH)_2$] is added to the blend to adsorb or hold fast to the polluting influences in the blend, and carbon dioxide (CO_2) gas is risen through the blend to hasten the lime as insoluble calcium carbonate precious stones. Lime ovens are utilized to create the CO_2 and lime utilized in carbonation; the lime is changed over to drain of lime in a lime slaker. The little, insoluble gems (created amid carbonation) settle out in a clarifier, after which the juice is again treated with CO2 (in the second carbonation tank) to evacuate the rest of the lime and polluting influences. The pH of the juice is bring down amid this second carbonation, causing huge, effectively filterable, calcium carbonate precious stones to shape. After filtration, a little measure of sulfur dioxide (SO2) is added to the juice to repress responses that prompt obscuring of the juice. Most offices buy SO as a fluid yet a couple of offices create SO2 by consuming essential sulfur in a sulfur stove. Following the expansion of SO2, the juice (known as thin squeeze) continues to the evaporators.

Before entering the next stage, the thin juice may receive soda ash to modify the pH and sulphitation with a sulfur-based compound to reduce colour formation due to decomposition of monosaccharides under heat.

4.6.4 Evaporation

The dissipation procedure, which expands the sucrose focus in the juice by expelling water, is regularly performed in a progression of five evaporators. Steam from huge boilers is utilized to warm the principal evaporator, and the steam from the water vanished in the main evaporator is utilized to warm the second evaporator. This exchange of warmth proceeds through the five evaporators, and as the temperature diminishes (because of warmth misfortune) from evaporator to evaporator, the weight inside each evaporator is likewise diminished, enabling the juice to bubble at the lower temperatures gave in each consequent evaporator. Some steam is discharged from the initial three evaporators, and this steam is utilized as a warmth hotspot for different process radiators all through the plant. After vanishing, the level of sucrose in the "thick squeeze" is 50-65 percent. Crystalline sugars, created later simultaneously, are added to the juice and broke down in the high melter. This blend is then sifted, yielding an unmistakable fluid known as standard alcohol, which continues to the crystallization task.

4.6.5 Crystallization

Sugar is solidified by low-temperature container bubbling. The standard alcohol is bubbled in vacuum dish until the point that it winds up supersaturated. To start gem arrangement, the alcohol is either "stunned" utilizing a little amount of powdered sugar or is "seeded" by including a blend of finely processed sugar and isopropyl liquor. The seed precious stones are precisely developed through control of the vacuum, temperature, feed-alcohol augmentations, and steam. At the point when the gems achieve the coveted size, the blend of alcohol and precious stones, known as massecuite or fillmass, is released to the blender. From the blender, the massecuite is filled highspeed centrifugals, in which the fluid is centrifuged into the external shell, and the gems are left in the internal radial bin. The sugar precious stones are then washed with unadulterated heated water and are sent to the granulator, which is a mix rotational drum dryer and cooler. A few offices have isolate sugar dryers and coolers, which are all things considered called granulators. The wash water, which contains a little amount of sucrose, is pumped to the vacuum searches for gold. In the wake of cooling, the sugar is screened and after that either bundled or put away in vast receptacles for future bundling. The fluid that was isolated from the sugar precious stones in the centrifugals is called syrup. This syrup fills in as feed alcohol for the "second bubbling" and is brought once more into the vacuum container alongside standard alcohol and reused wash water. The procedure is rehashed once, bringing about the creation of molasses, which can be further desugarized utilizing a particle trade process called profound molasses desugarization. Molasses that isn't desugarized can be utilized in the creation of domesticated animals feed or for different purposes.

4.7 Testing for Sucrose in cane sugar (Fehling's Test)

In this test the presence of aldehydes but not ketones is detected by reduction of the deep blue solution of copper(II) to a red precipitate of insoluble copper oxide. The test is commonly used for reducing sugars but is known to be NOT

specific for aldehydes. For example, fructose gives a positive test with Fehling's solution as does acetoin.

Two solutions are required:

1. Fehling's "A" uses 7 g $CuSO_4.5H_2O$ dissolved in distilled water containing 2 drops of dilute sulfuric acid.
2. Fehling's "B" uses 35g of potassium tartrate and 12g of NaOH in 100 ml of distilled water. These two solutions should be stoppered and stored until needed.

For the test:

   i)   Mix 15 ml of solution-"A" with 15 ml of solution-"B"

ii) Add 2 ml of this mixture to an empty test tube.

iii) Add 3 drops of the compound to be tested to the tube.

iv) Place the tube in a water-bath at 60° C.

A positive test is indicated by a green suspension and a red precipitate. The test is sensitive enough that even 1mg of glucose will produce the characteristic red colour of the compound

CHAPTER V: LEATHER INDUSTRY

5.   Introduction

The skin of creature comprises of the epidermis, the dermis, and subcutaneous tissue. For the produce of cowhide just the dermis is imperative. A creature skin is dealt with for human utilize. The fundamental crude materials for the calfskin business are cowhide from dairy cattle and other domesticated animal's creatures, buckskin, crocodile skin and snake skin. All are utilized for shoes, garments and other design embellishments. Calfskin is likewise utilized in upholstery, inside improving, horse tack and outfits. Such skins are once in a while still assembled from chasing and handled at a household or distinctive level yet most cowhide making is currently industrialized and extensive scale. Different tannins are utilized for this reason.

5.1 Preparation of Skin for tanning

Tanning is the way toward changing over crude stows away or creature skins into cowhide. Covers up and skins can assimilate tannic corrosive and other synthetic substances that keep them from rotting, make them impervious to wetting, and keep them supple and solid. How this procedure is done importantly affects the characteristics of the pack you want to use for a considerable length of time to come. Once a creature skin has been transformed into a cover up by being degreased and having all its hair evacuated, the tanning procedure can start. The surface of skins contains the hair and is known as the grain side. The substance side of the cover up or skin is significantly thicker and milder. An untreated skin would both solidify and rot as it is a natural material. Thus, the point of the tanning procedure is to keep this from happening – to transform the stow away into cowhide.

The essential guideline has been the same for every one of these centuries: To change the protein called collagen, which the skin is comprised of. You can really get a feeling of this protein with the stripped eye. Collagen particles jump at the chance to initially arrange and afterward to bend together into "fiber packages" that you can without much of a stretch check whether you look carefully enough at great quality calfskin.

The fundamental preparing arrangements of cowhide make. Where skins are handled with the hair or fleece on, the unhairing and liming forms are precluded and supplanted by a scouring (washing) of the fleece or hair. The primary compound procedures completed by the leather expert are the unhairing, liming, tanning, killing and coloring.

Step 1 - Unhairing

The keratin of hair and fleece is liable to assault by soluble base, which will break (by hydrolysis) the sulfur-sulfur bond in the cystine linkage of the keratin. The hair is assaulted first at the root, where it is in its youthful shape. After a specific time of being saturated with an answer of soluble base (sodium or calcium hydroxide) and a diminishing specialist, typically sodium sulfide, the hair roots are disintegrated, and the hair might be expelled and spared.

Step 2 Liming

The way toward liming is a blend of substance and physical activity on the skin structure. By including lime and sulfur intensify the hair is expelled from the skin. Liming is the most widely recognized technique for hair evacuation, yet warm, oxidative, and compound strategies likewise exist. An answer of lime and sodium sulfide is utilized and the skins are inundated in this answer for whatever time is important to create the coveted effects.The ordinary methodology for liming is to utilize a progression of pits or drums containing lime mixers (calcium hydroxide). The procedure of unhairing is taken to fulfillment amid the liming procedure, and there is calculable adjustment of the collagen because of the activity of antacid.

Skin protein (collagen):

1) Hydrolysis of amide groups (propylamide)

$$R-\underset{\underset{O}{\|}}{C}-NH_2 + OH^- \longrightarrow R-\underset{\underset{O}{\|}}{C}-O^- + NH_3$$

2) Modification of guanide groups

$$R-(CH_2)_3-NH-C\underset{\diagdown NH}{\overset{\diagup NH_2}{}} + H_2O + OH^- \longrightarrow R-(CH_2)_3-NH-\underset{\underset{O}{\|}}{C}-O^- + 2NH_3$$

3) Hydrolysis of keto-imide links in protein chains

$$-CH_2-\underset{\underset{O}{\|}}{C}-NH-CH_2- + OH^- \longrightarrow -CH_2-\underset{\underset{O}{\|}}{C}-O^- + H_2N-CH_2-$$

4) Swelling: Notwithstanding the compound activity, within the sight of a salt, swelling of the stringy structure happens. This is because of an osmotic weight impact. The outcome is the detachment of the filaments and the fibrils from each other and an opening up of the entire structure.

5) Removal of unwanted material: Present in the collagen structure are globular proteins and other interfibrillary substances. These are hydrolysed and evacuated in the ensuing washing and bateing forms. Undesirable fats are saponified.

Step 3 - Deliming and Bateing

After the solid basic activity, the skin structure is additionally opened up amid the deliming and bateing process. Hydrolysis is proceeded by the chemical procedure and further disintegration of undesirable material happens. The primary sinewy system is then tidied up and the swelling is lessened. Bateing is completed at pH 9-10. This is accomplished by treatment with corrosive salts (ammonium chloride or ammonium sulfate) or carbon dioxide until the point that the coveted pH is reached.It is an enzymatic procedure performed to bestow delicate quality, stretch, and adaptability to calfskin. This progression can take between 30 minutes and 12 hours. Bating and deliming are generally performed together by setting the stows away in fluid arrangement of ammonium salt and proteolytic compounds at 27 °C to 32 °C.

Step 4 - Pickling

Pickling is the procedure of synthetically expulsion of undesirable material from the skin to acquire clean surface. The principle fermentation process is that of pickling. The skins are disturbed in an answer of salt also, sulphuric corrosive until the point when they are at or close harmony at a pH estimation of 3.0 - 3.5.after Pickling a washing procedure happens taken after by inundation in a lime-water shower to kill any staying corrosive.

Step 5 - Tanning

Vegetable tanning: Vegetable-tanned cowhide is tanned utilizing tanning and different fixings found in vegetable issue, tree covering, oak, chestnut or mimosa, however many tree composes and different plants have been utilized. For by far most of the previous a huge number of years, this adjustment has been performed by absorbing it an answer made up of vegetable tannins. Overwhelming calfskins and sole cowhides are delivered by the vegetable tanning process, the most seasoned of any procedure being used in the calfskin tanning industry. Truth be told, the word 'tannin' gets from an old German word for 'fir'. So 'tanning' has nothing to do with shading as in getting your unit off and giving the sun a chance to dark colored your skin. Vegetable-tanned cowhide isn't steady in water; it has a tendency to stain, and if left to splash and after that dry it will contract and turn out to be less supple and harder.

A tannin is an atom that bonds effectively with proteins and will coax fluids out. On the off chance that you are a red wine consumer, you may have had warmed discussions over a wine's 'tannins' – the fixing that influences the wine to feel dry in the mouth, some of the time to the degree of making both your tongue and gums feel disagreeably dry and fruitless. Similarly as the tannins in wine originate from the skin of the grape, so the tannins in trees are found in the bark.

When tanning covers up to make cowhide, the stows away are absorbed a tanning arrangement. The tannin particles will enter the cover up and dislodge the water that is bound in the collagen. The water is drawn out, however as the tannins replace the evacuated water, the calfskin does not become unyielding as completely got dried out cowhide generally would.

It might sound simple, yet it isn't. The procedure is mind boggling and the skins require numerous medicines over a time of up to two months all together for the water atoms to be completely extricated and giving the tannin particles a chance to take their places in simply the correct way. A great deal of work from exceptionally talented specialists is included as well.

Mineral (chromium) tanning: The many-sided quality, cost and time required with tanning with vegetable tannins drove, in 1858, to the advancement of utilizing mineral tanning specialists. It is more supple and flexible than vegetable-tanned calfskin, and does not stain or lose shape as radically in water as vegetable tanned. Otherwise called wet-blue for its shading got from the chromium. Chrome tanning is performed in light of the response between the stow away and a trivalent chromium salt, for the most part a fundamental chromium sulfate and it is the most prevalent mineral tanning operator today. The entire procedure can be computerized and completed in multi day, and the chrome particles uprooting the water and official with the collagen are significantly littler than vegetable tanning atoms. This by and large makes chrome tanned calfskin more slender and milder than vegetable tanned cowhide.

The procedure, be that as it may, is less normal than when utilizing vegetable tannins. It includes first setting the stows away in acidic salts to more readily make the chrome fit in the middle of the collagen atoms – and after that restoring the covers up to an ordinary pH level.

This requires the utilization of acids and different synthetics and the chromium sulfates themselves. If not legitimately dealt with, these will have a negative ecological effect, and the business keeps on being feeling the squeeze to "tidy up" as more directions are presented.

Step 6 - Leather finishing

Completing Leathers might be done in an assortment of courses: buffed with fine abrasives to deliver a softened cowhide complete; waxed, shellacked, or treated with shades, colors, and saps to accomplish a smooth, cleaned surface and the coveted shading; or lacquered with urethane for a gleaming patent calfskin. Completing comprises of applying a surface covering - shades or colors bound in a natural (acrylic, butadiene or polyurethane) or protein (casein) medium. This upgrades the regular characteristics of the skin and covers such imperfections as scars, horn harm, seed scars and so on., as might be available. The principle prerequisites for completing are eveness and the reproducibility of shading and satisfactory wear and feel properties.

CHAPTER VII: CHEMICAL FOOD STAFF PROCESSING

6.   Introduction

Fermentation is the chemical break down of substances by bacteria, yeast, or other microorganisms, especially that involved in the making of beers, wines and spirits in which sugar are converted to ethyl alcohol. Fermentation is a reaction where in a raw material is converted into a product by the action of micro-organisms or by means of enzymes. When micro-organisms are used, they produce enzymes *in-situ* which then catalyzes fermentation reactions. Fermentation is also used more broadly to refer to the bulk growth of microorganisms on a growth medium, often with the goal of producing a specific chemical product. French microbiologist Louis Pasteur is often remembered for his insights into fermentation and its microbial causes. The science of fermentation is known as zymology.

Fermentation takes place when the electron transport chain is unusable (often due to lack of a final electron receptor, such as oxygen). In this case it becomes the cell's primary means of ATP (energy) production. It turns Nicotinamide adenine dinucleotide (NADH) and pyruvate produced in glycolysis into NAD^+ and an organic molecule (which varies depending on the type of fermentation; see examples below). In the presence of O_2, NADH and pyruvate are used to generate ATP in respiration. This is called oxidative phosphorylation, and it generates much more ATP than glycolysis alone. For that reason, cells generally benefit from avoiding fermentation when oxygen is available, the exception being obligate anaerobes which cannot tolerate oxygen.

The first step, glycolysis, is common to all fermentation pathways:

$$C_6H_{12}O_6 + 2\,NAD^+ + 2\,ADP + 2\,P_i \rightarrow 2\,CH_3COCOO^- + 2\,NADH + 2\,ATP + 2\,H_2O + 2H^+$$

Pyruvate is $CH_3COCOO-$. Pi is inorganic phosphate. Two ADP molecules and two Pi are converted to two ATP and two water molecules via substrate-level phosphorylation. Two molecules of NAD^+ are also reduced to NADH.

6.1 Alcohol beverages

An alcoholic beverage is a drink containing ethanol, commonly known as **alcohol**. Alcoholic beverages are divided into three general classes: beers, wines, and spirits, that contains ethyl alcohol, or ethanol (CH_3CH_2OH), as an intoxicating agent. Alcoholic beverages are fermented from the sugars in fruits, berries, grains, and such other ingredients as plant saps, tubers, honey, and milk and may be distilled to reduce the original watery liquid to a liquid of much greater alcoholic strength.

Beer is an alcoholic drink made from yeast-fermented flavoured with hops. It is also made from malt, corn, rice, and hops. Beer is the best-known member of the malt family of alcoholic beverages, which also includes ale, stout, porter, and malt liquor. Beers range in alcoholic content from about 2 percent to about 8 percent. Wine is made by maturing the juices of grapes or different organic products, for example, apples (juice), fruits, berries, or plums. Winemaking starts with the collect of the natural product, the juice of which is aged in extensive vats under thorough temperature control. At the point when aging is finished, the blend is sifted, matured, and packaged. Normal, or unfortified, grape wines for the most part contain from 8 to 14 percent liquor; these incorporate such wines as Bordeaux, Burgundy, Chianti, and Sauterne. Strengthened wines, to which liquor or schnaps has been included, contain 18 to 21 percent liquor; such wines incorporate sherry, port, and muscatel.

The making of refined spirits starts with the pounds of grains, natural products, or different fixings. The resultant matured fluid is warmed until the point when the liquor and flavorings vaporize and can be drawn off, cooled, and dense again into a fluid. Water stays behind and is disposed of. The concentrated fluid, called refined refreshment, incorporates such alcohols as bourbon, gin, vodka, rum, liquor, and mixers, or cordials. They run in alcoholic substance as a rule from 40 to 50 percent, however higher or bring down fixations are found.

In the ingestion of a mixed refreshment, the liquor is quickly invested in the gastrointestinal tract (stomach and digestion tracts) since it doesn't experience any stomach related procedures; accordingly, liquor ascends to abnormal states in the blood in a moderately brief time. From the blood the liquor is circulated to all parts of the body and has a particularly articulated impact on the cerebrum, on which it applies a depressant activity. Affected by liquor the elements of the cerebrum are discouraged in a trademark design. The most complex activities of the mind—judgment, self-feedback, the restraints gained from soonest adolescence—are discouraged first, and the loss of this control results in a sentiment of energy in the beginning times. Hence, liquor is here and there thought of, incorrectly, as a stimulant. Affected by expanding measures of liquor, the consumer slowly turns out to be less alarm, attention to his condition winds up diminishes and murky, solid coordination disintegrates, and rest is encouraged.

6.2 Manufacture of Beer

Step 1 Ingredient: Beer is made from four basic ingredients: Barley, water, hops and yeast. It is an alcoholic drink made from yeast-fermented malt flavoured in hope. The basic idea is to extract the sugars from grains (usually barley) so that the yeast can turn it into alcohol and CO_2, creating beer.

Step 2 Malting: The brewing process starts with grains, usually barley (although sometimes wheat, rye or other such things.) It is the process of germination of barley under controlled conditions. Varying conditions during the malting process (temperature and humidity) allows different types of malt to be obtained, giving different colours and flavours to the beer and involves the process that liberates the starch-converting enzymes that are naturally present in the barley grain. The grains are harvested and processed through a process of heating, drying out and cracking. The main goal of malting is to isolate the enzymes needed for brewing so that it's ready for the next step.

Step 2 Malting

Is the controlled germination of cereal grains, mostly barely in the moist air followed by the controlled drying of the grains. It is the process in which barely are treated to convert the insoluble starch to soluble sugars. In malting simplifying proteins, generating nutrients for yeast and the development of enzymes will also be there. During malting large molecular weight content of the endosperm cells walls, the storage proteins and the starch granules are hydrolyzed by enzyme rendering them more soluble in water. The purpose of malting include: to increase the enzymes activity for endosperm modification, characteristic flavor with a minimum loss of dry weight. During malting the wet process is start with steeping to germination start and ends with kilning which reduces the moisture and produces a stable final product.

Step 3 Mashing: The flour from the cereals (malt and other unmalted cereals) is mixed with water and subjected to certain processes to obtain a wort of a suitable composition for the kind of beer being produced (varying times, temperatures and PH) or Mashing is the process of combining a mix of combined grain (malted barley) and water known as liquor and heating this mixture it allows the enzymes in the malt to break down the starch in the grain into sugar, typically maltose to create a malty liquid called wort. Wort is the sweet infusion of ground malt or other grain before fermentation, used to produce beer and distilled malt liquor. Mashing lasts 2 to 4 hours and finishes with a temperature of approximately 75 °C.

Step 4 Boiling: The diluted and filtered wort is boiled for about an hour while hops and other spices are added several times. Hops are the small, green cone-like fruit of a vine plant. They provide bitterness to balance out all the sugar in the wort and provide flavor. They also act as a natural preservative, which is what they were first used for. The purpose of boiling wort is to: Transform and make soluble the bitter substances in the hops, Eliminate undesirable volatile substances, Sterilise the wort, establish the final concentration of wort.

Step 5 Fermentation: Once the hour long bubble is over the wort is cooled, stressed and separated. It's at that point put in an aging vessel and yeast is added to it. Now the blending is finished and the maturation starts. The brew is put away for two or three weeks at room temperature (on account of lagers) or numerous weeks at chilly temperatures (on account of ales) while the yeast works its maturation enchantment. Essentially the yeast gobbles up everything that sugar in the wort and releases CO_2 and liquor as waste items.

Step 6 Bottling and Aging: You've now got jazzed up brew, anyway it is still level and uncarbonated. The level lager is packaged, at which time it is either falsely carbonated like a pop, or if it will be 'bottle adapted' it's permitted to normally carbonate through the CO_2 the yeast produces. In the wake of enabling it to age for anyplace from half a month to a couple of months.

6.3 Manufacture of sprite

Distilled spirits are produced from agricultural raw materials such as grapes, other fruit, sugar-cane, molasses, potatoes, cereals, etc.

There are many subtleties involved in the creation of different spirits drinks but, by way of example, the process for a cereal-based spirits is as follows:

Step 1: Milling. The raw material is ground into a coarse meal. The process breaks down the protective hull covering the raw material and frees starch.

Step 2: Mashing. The starch is converted to sugar, which is mixed with pure water and cooked. This produces a mash.

Step 3: Fermentation. The sugar is converted to alcohol and carbon dioxide by the addition of yeast. With the addition of yeast to the sugar, the yeast multiplies producing carbon dioxide which bubbles away and a mixture of alcohol, particles and congeners, or the elements which create flavour to each drink.

Step 4: Distillation. The alcohol, grain particles, water and congeners are heated. The alcohol vaporises first, leaving the water, the grain particles and some of the congeners in the boiling vessel. The vaporised alcohol is then cooled or condensed, to form clear drops of distilled spirits.

Two additional steps are often taken in making some distilled spirits –

Step 5: Ageing. Certain distilled spirits (e.g. rum, brandy, whisk(e)y) are matured in wooden casks where they gradually develop a distinctive taste, aroma and colour.

Step 6: Blending. Some spirits go through a blending process whereby two or more spirits of the same category are combined. This process is distinctive from mixing since the blended spirit remains of the same specific category as its components.

6.4 Manufacture of wines

Wine making has been around for a large number of years. It isn't just a craftsmanship yet additionally a science. Wine making is a characteristic procedure that requires minimal human intercession, however each wine creator controls the procedure through various methods. When all is said in done, there are five fundamental segments of the wine making process: reaping, pulverizing and squeezing, aging, illumination, and maturing and packaging. Wine creators commonly take after these five stages however include varieties and deviations en route to make their wine remarkable.

Step 1 Harvesting:

In the generation procedure of wine the initial step is called Harvesting and it is a vital piece of guaranteeing scrumptious wine. Grapes are a green, purple or dark berry developing in groups on vine, eaten as foods grown from the ground in making wine. The grapes are picked when they are ready, for the most part as controlled by taste and sugar readings. Grapes are the main organic products that have the vital acids, esters, and tannins to reliably make regular and stable wine. Tannins are textural components that make the wine dry and add severity and astringency to the wine

The minute the grapes are picked decides the sharpness, sweetness, and kind of the wine. Deciding when to reap requires a dash of science alongside antiquated tasting. The acridity and sweetness of the grapes ought to be in idealize adjust, however reaping additionally vigorously relies upon the climate.

Gathering should be possible by hand or mechanically. Many wine creators like to collect by hand in light of the fact that mechanical gathering can pummel the grapes and the vineyard. Once the grapes are taken to the winery, they are arranged into bundles, and spoiled or under ready grapes are expelled.

Step 2 Crushing and Pressing: After the grapes are arranged, they are prepared to be de-stemmed and pulverized. For a long time, people did this physically by stepping the grapes with their feet. These days, most wine producers play out this mechanically. Mechanical presses step or trod the grapes into what is called must. Must is essentially crisply squeezed grape squeeze that contains the skins, seeds, and solids. Mechanical squeezing has brought enormous clean gain and additionally expanded the life span and nature of the wine.

For white wine, the wine creator will rapidly pulverize and press the grapes keeping in mind the end goal to isolate the juice from the skins, seeds, and solids. This is to keep undesirable shading and tannins from draining into the wine. Red wine, then again, is left in contact with the skins to procure flavor, shading, and extra tannins.

Step 3 Clarification: When maturation is finished, elucidation starts. Elucidation is the procedure in which solids, for example, dead yeast cells, tannins, and proteins are evacuated. Wine is exchanged or "racked" into an alternate vessel, for example, an oak barrel or a treated steel tank. Wine would then be able to be illuminated through fining or filtration.

Fining happens when substances are added to the wine to illuminate it. For instance, a wine producer may include a substance, for example, dirt that the undesirable particles will hold fast to. This will constrain them to the base of the tank. Filtration happens by utilizing a channel to catch the bigger particles in the wine. The illuminated wine is then racked into another vessel and arranged for packaging or future maturing.

Step 4 Aging and Bottling

Maturing and packaging is the last phase of the wine making process. This is done precisely so the wine does not interact with air. Better wines might be put away for quite a long while in bottles before they are discharged. A wine producer has two choices: bottle the wine immediately or give the wine extra maturing. Additionally, maturing should be possible in the containers, hardened steel tanks, or oak barrels. Maturing the wine in oak barrels will create a smoother, rounder, and more vanilla enhanced wine. It likewise builds wine's presentation to oxygen while it ages, which diminishes tannin and enables the wine to achieve its ideal fruitiness. Steel tanks are usually utilized for lively white wines

www.ingramcontent.com/pod-product-compliance
Lightning Source LLC
Chambersburg PA
CBHW021016180526
45163CB00005B/1980